# The Topology of the 2×2 Games

*The Topology of 2×2 Games* provides the reader with a unique new guide to this basic tenet of microeconomics. David Robinson and David Goforth examine the structure of players' preferences and the resulting topology to create a greater awareness of these games and the relationships that exist between them.

This systematic treatment of 2×2 games – presented in tabular format – will provide researchers and teachers in game theory and microeconomics with an indispensable tool.

**David Robinson** is Associate Professor of Economics, and **David Goforth** is Assistant Professor of Computer Science. They are both at Laurentian University, Canada.

**Routledge Advances in Game Theory**
Edited by Christian Schmidt

**Volume 1**
**Game Theory and Economic Analysis**
A Quiet Revolution in Economics
*Christian Schmidt*

**Volume 2**
**Negotiation Games**
Applying Game Theory to Bargaining and Arbitration
*Steven J. Brams*

**Volume 3**
**The Topology of the 2×2 Games**
A New Periodic Table
*David Robinson and David Goforth*

# The Topology of the 2×2 Games

## A New Periodic Table

**David Robinson and David Goforth**

Routledge
Taylor & Francis Group
LONDON AND NEW YORK

First published 2005 by Routledge
2 Park Square, Milton Park, Abingdon, Oxon OX14 4RN

Simultaneously published in the USA and Canada
by Routledge
270 Madison Ave, New York, NY 10016

*Routledge is an imprint of the Taylor & Francis Group*



Printed and bound in Great Britain by
TJ International Ltd, Padstow, Cornwall



*British Library Cataloguing in Publication Data*
A catalogue record for this book is available from the British Library

*Library of Congress Cataloging in Publication Data*
A catolog record for this title has been requested

ISBN 0–415–33609–0 (hbk)

# Contents

# List of Figures

# List of Tables

# A New Foundation

In *The Topology of the* $2 \times 2$ *Games* we jack up the entire edifice of $2 \times 2$ game theory and put a foundation under it. Our target audience is the next generation of teachers and researchers, and our goal is to provide a systematic approach to the $2 \times 2$ games that is both easy to learn and powerful.

The core of the book is the Periodic Table of the $2 \times 2$ Games in Chapter 9. It lays out relationships among all 144 $2 \times 2$ strict ordinal games in a format that is very like the Periodic Table of Elements. If we have been successful, the table will be an indispensable tool for researchers and teachers in any field that uses game theory.

The structure of the table is a direct result of the structure of the players' preferences. Preferences induce a topology on the $2 \times 2$ games, hence the name of the book. The topology yields new information about the $2 \times 2$ games. Chapter 5, for example, investigates the Prisoner's Dilemma(PD). The Prisoner's Dilemma occupies a unusual, even central position among the $2 \times 2$ games. It lies at a kind of crossroads in the topological space. Exploring the neighbourhood of the Prisoner's Dilemma allows us to identify asymmetric versions of the PD we call Alibi Games. With the PD they make up the *Prisoner's Dilemma Family* (PDF). The Prisoner's Dilemma Family in turn forms the boundary of the class of games we identify with the social dilemmas.

Symmetries in the topological space of the $2 \times 2$ games also yield new results. In Chapter 8 we present a new approach to what Schelling [33] called the "mixed motive" games. We introduce what might be called a *topography* of the games based the degree of conflict implicit in the payoff structure.

Mathematically inclined readers should be warned. The book is really about $2 \times 2$ games, not topology. Topology is essential to the analysis, but only those elements that make the $2 \times 2$ games understandable are introduced. Concepts from graph theory and the theory of groups are also introduced where they provide useful insights about the structure of the $2 \times 2$ games. Readers new to topology and group theory will find that, almost in passing, they have established a beachhead for further study in these topics, as well as practical tools that make game theory more accessible.

And they do make the $2 \times 2$ games more accessible. We have been pleasantly surprised to find that second-year economics students can quickly learn to apply the model to construct and to describe all the most famous games using the expositional devices we introduce.

The fact that the topology of the $2 \times 2$ games is so simple and elegant raises a question. Why was it not discovered much earlier? The reason, we suspect, is simply that pioneers in the new field of game theory built very quickly. Von Neumann and Morgenstern [39] laid down a very solid foundation, but the floor space set aside for the $2 \times 2$ games turned out to be inadequate. New rooms were attached to the main structure in an altogether haphazard manner.

In an attempt to provide some order, Rapoport and Guyer [23], Rapoport, Guyer and Gordon [24] and Brams [5] produced *typologies* of the $2 \times 2$ games. Typologies, however, belong to an early stage of science, to be replaced as soon as the deeper relationships are understood. Members of the plant kingdom, for example, might be seen as falling into one of two types – big and small. Under this typology, bamboo is of the same type as the alder tree. In evolutionary terms, bamboo is actually closer to wheat than to alder. Botany long ago abandoned typologies in favour of the "phyletic" approach based on a strict concept of what it means to be related.

To get beyond the typological approach requires a notion of what it means for games to be related. It turns out that preferences provide the appropriate notion of closeness. The topology induced by preferences is beautiful and it makes the systematic treatment of the $2 \times 2$ games possible.

## The organization of the book

Chapter 1 begins with a brief introduction to game theory. The main innovation in Chapter 2 is to shift the analysis from the space of strategies to the space of payoffs. This dual to the familiar matrix representation provides an intuitive introduction to the $2 \times 2$ games, but it also provides a convenient foundation for the topological analysis of Chapter 3. The *order graph* introduced here allows us to present and compare games in payoff space.

Order graphs rely on the *inducement correspondence* for the Nash situation, a powerful concept developed by Greenberg [10]. The inducement correspondence is particularly suited to the payoff-space representation.

We also present a simple indexing system based on the topological relationships between games. Chapter 3 introduces one of the basic reference tools, a set of four figures show order graphs for all 144 games arrayed according to our indexing system. The figures are the basis of the Periodic Table of the $2 \times 2$ Games developed in Chapter 9.

In Chapter 3 we show how preferences induce the topology on the set of games and generate a graph with 144 nodes and 432 edges. In the graph, games that are related economically are *near* each each other topologically. Investigating the remarkable regularities and symmetries of this structure is the main enterprise of this book. Because the simplest language for describing the subspaces and the subgraphs comes from group theory, we introduce several useful terms and concepts from graph theory.

Chapters 4 to 8 explore specific topological subspaces. Proper subspaces typically contain games that are related in an economically interesting way. For example, seven of the best known games are in a group of 12 symmetric games picked out by applying the "symmetric operators". Chapter 4 examines this 12-game subspace and goes on to identify symmetries, rotations and reflections, including the special sense of reflection used by previous authors. Chapter 5 examines the Prisoner's Dilemma and identifies a class of related games we call the *Alibi games*.

Chapters 6 and 7 examine pipes and *hotspots*, the most peculiar topological features of the space of the $2 \times 2$ games. These chapters reveal that the map of the $2 \times 2$ games cannot be embedded in a surface with fewer than 37 holes. This may be the most peculiar and useless fact in the entire book. It is closely related to the fact that the Periodic Table of the $2 \times 2$ games is considerably more complex than the Periodic Table of the Chemical Elements. The hotspots and pipes are like wormholes that link regions of the periodic table through other dimensions.

There are no zero-sum games among the strict ordinal games, but there are constant rank-sum games, both with and without Nash equilibria. There are also games of pure conflict that are not constant rank-sum and games that have been called "no-conflict" games [23] [24][5]. Chapter 8 treats the games of pure conflict, pure common-interest and mixed motives systematically and proposes a new category, the *Type games*.

Chapter 10 shows how the topological approach can be applied in a continuous space. Results from evolutionary experiments in the continuous subspace of the symmetric games show that, while the topological relationships continue to hold for real-valued games, the boundaries of the ordinal games are not always the relevant behavioral boundaries.

Topology lends itself to a diagrammatic exposition. The 94 figures provide a flexible system for analysing the $2 \times 2$ games.

# Chapter 1

# 2×2 games and the strategic form

The 2 × 2 games are usually the first that students meet and probably the last they forget. Special cases, like the Prisoner's Dilemma, Chicken, Coordination game and the Battle of the Sexes, are the most familiar formal descriptions of social situations in all of the social sciences. Specific examples are routinely discussed in philosophy, biology, law, sociology, politics and every other field in which strategic situations arise.

Simplicity gives the 2 × 2 games their power: they provide remarkable diversity with the absolute minimum of machinery. The strategic situation involves only two players, each with only two alternatives. There are only four possible outcomes and each outcome is described by a single payoff for each player. A game is therefore fully described by just 8 numbers.

The apparent simplicity of the 2 × 2 games is deceptive. The eight numbers yield a class of 144 problems of remarkable richness and complexity. And while individual games have been discussed in detail, the relationships among the games have never been mapped. Instead, the 2 × 2 games are almost always dealt with anecdotally.

The goal of this book is to present a systematic framework for the 2 × 2 games. In this chapter we present the standard representation and discuss some of the core concepts in game theory. Nothing in this chapter should be new for readers familiar with game theory. For

| Form | common representation | mapping from | to payoffs for |
|---|---|---|---|
| Strategic | matrix | combinations of plans | persons |
| Extensive | tree | contingency plans | persons |
| Characteristic Function | list | possible coalitions | coalitions or persons |

Table 1.1: Standard forms

others the chapter will provide a useful overview and an introduction to the conventions for describing games.

## 1.1 Form and solution

Abstracting vital information and suppressing the irrelevant is at the heart of any formal approach. Game theory is a formal approach to analysing social situations employing highly stylized and parsimonious descriptions.

### Form

One important and standardized block of information in the formal descriptions used by game theorists is called a game *form*. A form specifies the *payoffs* associated with every possible combination of decisions. There are several widely used forms, including the *strategic form*, typically presented in a matrix, the *extensive form*, which is usually represented as a tree, and the *characteristic function form*, expressed as a function on subsets of players.

The form is a minimal representation of a social situation. It is almost always supplemented with variable elements that fill out the description. These elements include specific rules of play and the timing of moves, descriptions of the information players have about the situation, and even of the opinions and thought processes of players. Figure 1.1 illustrates the way elements are introduced on top of the basic form.

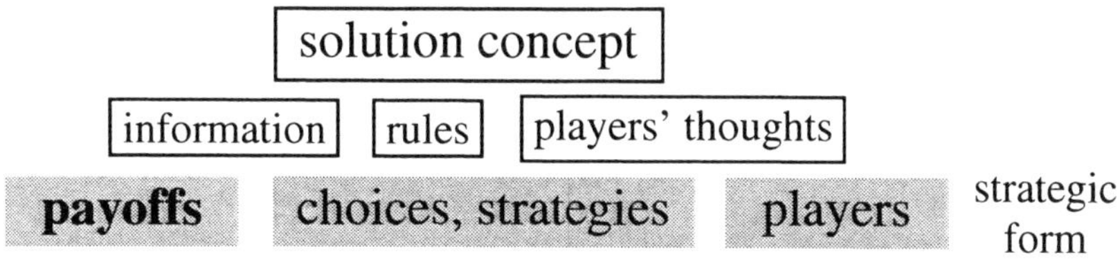


Figure 1.1: Elements of a game model

### Solution

The analysis of a game is usually directed toward determining what might happen when players interact given the rules and available information. "What might happen" is the *solution*. A solution describes both the actions of the players and the payoffs that result[1].

The solution for a particular situation is "picked out" by a *solution concept*, which is ultimately a statement about what matters for the players. An example may help. Imagine that players only care about avoiding low payoffs. Players like this would identify the worst payoff associated with each alternative and then choose the alternative with the *best* worst outcome. Attempting to maximize the minimum is a feature of the players' behaviour and ultimately of their thinking. To say that the outcome is determined by such thinking is to impose a solution concept.

A solution concept allows us to read *from* the payoffs that are possible in a given situation *to* the actions that would be chosen by players of a certain type. It generally yields a subset of the possible payoffs as a solution. Occasionally a sensible solution concept will select an outcome that seems altogether unacceptable, or even fail to select an outcome. Payoff patterns that produce problems for sensible solution concepts are especially interesting. Multiple solutions,

[1]For Von Neumann and Morgenstern "The immediate concept of a solution is plausibly a set of rules for each participant which tell him how to behave in every situation which may conceivably arise." They go on to call the pattern of payoffs resulting from the play of a game an "imputation" and identify the imputation, if it exists with the solution. Since a unique imputation does not always exist, they note that the "notion of a solution will have to be broadened considerably" ([39] p.34).

non-existent solutions and unacceptable solutions all occur among the 2 × 2 games. Furthermore, these problematic payoff patterns appear to describe real situations.

Our analysis focuses on the payoff structure rather than the behaviours or solution concepts. We stop where most analysis begins. This restriction is less limiting than it might seem because players are motivated by payoffs, and their thoughts about how to play must be related in some systematic way to the structure of payoffs. Any solution concept, similarly, has to relate behaviour to the pattern of payoffs. Nothing can be extracted that is not already implicit in the form. The form of a game fills a role rather like the set of axioms and rules for reasoning in Euclidean geometry. What is extracted from the form, however, may depend on features of the situation that are not part of the form itself.

## 1.2 2×2 games in strategic form

A game in strategic form is just a function with one input for each player (a strategy) and one output for each player (a payoff). More formally, a game in strategic form is a vector function and its domain, the *strategy space*. The strategy space is just the set of all possible combinations of strategies, and therefore incorporates both the player and strategy sets.

The ordinal 2 × 2 games are the simplest of all games in strategic form, with only two *players*, both of whom choose once between the two actions available to them. A player without at least two alternatives has no choice and his or her decisions cannot matter. A situation with less than two players that matter has no strategic interactions and is not a game.

The two players constitute the *player set*. The actions available to the players, called *strategies*, make up the *strategy sets* for the players. Individual strategies may be as simple as the selection of a destination from a signpost or as complex as Napoleon's battle plan for the conquest of Russia. Because we are interested in the payoff function, we can suppress detailed information about the chain of actions that make up a particular strategy.

The expression "2×2 *game*" is simply a description of the strategy space. It says that the strategy space is the cross-product of two strategy sets, each with exactly two alternatives. There are therefore *four* strategy combinations. Similarly, a 3 × 4 game is a two-player game in which one player has three alternatives and the other has four. In the 3 × 4 game there are 12 possible outcomes. A 2 × 2 × 2 game has three players, each with two alternatives.

## Equivalent games

With four outcomes and two players, a 2 × 2 game is completely described by eight numbers. An array with eight numbers is just an address in an 8-dimensional Cartesian payoff space, and there are uncountably many 2 × 2 games, each fully described by an 8-number address. One goal of this book is to present a useful way to divide this infinite 8-space into a manageable number of meaningful regions.

The regions are defined using two simplifications: (i) we treat sets of games that are equivalent under strictly monotonic transformations as equivalence classes, and (ii) we rule out indifference.

Any game in one of the classes can be converted into any other in the same region by some strictly monotonic transformation. Since a monotonic transform conserves order, all the games in an equivalence class are ordinally equivalent. These equivalence classes partition the 8-dimensional payoff space for the 2 × 2 games into 144 regions. An *ordinal 2×2 game* is a 2 × 2 game with a payoff function that maps from the strategy space to these equivalence classes.

Ruling out indifference eliminates ties in the payoffs for either player, restricting us to the *strict* ordinal games. It also creates a discontinuity between the equivalence classes[2].

[2]Each region is an open set, since games with outcomes that are not strictly ranked – games with ties – can occur only as the limit of a monotonic transform of a game with strictly ranked outcomes. They appear in the space of measure zero between regions.

| | | Column Player's Strategies | |
|---|---|---|---|
| | | *L* | *R* |
| *Row Player's Strategies* | *U* | 1,4 | 3,3 |
| | *D* | 2,2 | 4,1 |

Table 1.2: Payoff matrix: standard notation for the strategic form

**Representative games**

Any four ordered elements will serve to represent the ordinal equivalence classes. In keeping with common practice we use payoffs constructed using 1, 2, 3 and 4 for each player. The resulting set of representative games is discrete subspace of the continuous payoff space. *The Topology of the* $2 \times 2$ *Games* is about the relationships among these representative games.

## 1.3 Conventions for payoff matrices

There is a remarkably economical notation for keeping track of which of the eight numbers is assigned to which player in which situation. A matrix like the one in Table 1.2 is often called the *payoff matrix*[3]. If we have the payoff matrix we have all the information for a game in strategic form [4].

The game illustrated in Table 1.2 is the Prisoner's Dilemma, the most famous game of all. Notice that the two players are named and that for each player two possible actions are identified. The names of the players and the strategies can be changed to suit the situation without affecting the nature of the game. The payoffs are those of the representative game. Any ordinally equivalent game is also a Prisoner's Dilemma.

It is convenient to name the row player "Row" and the column

[3]The matrix of vectors describing a 2-person game is often called a *bi-matrix* because it can be written as two separate payoff matrices, one for each player.

[4]Von Neumann and Morgenstern called this the normal form but strategic form is more descriptive and is now preferred by most writers.

player "Column". In the game described in Table 1.2, Row can choose either $U$ or $D$. Column can choose $L$ or $R$. If Row were to make a commitment to choosing $U$, then Column would only need to consider the payoffs in the row labelled $U$. Having made such a commitment, Row would still not know the outcome of the game unless she could predict Column's decision.

A *strategy combination* (often called a "strategy profile") identifies a possible outcome. For example, if Row chooses $U$ and Column chooses $L$, the payoffs for the two players are given in the row labelled $U$ and the column labelled $L$. For a game in matrix form, a strategy combination is a kind of "matrix address". It is conventional to write strategy combinations in parentheses, with the row player's strategy first: $(U,L)$. The same convention is used for writing payoffs, so the payoff pair $(1,4)$ tells us that Row gets 1 if the strategy combination $(U,L)$ is selected and Column gets 4.

We usually think of the 1 and 4 as representing utility or some generalized measure of joy, but the numbers could represent pesos, dollars or quantities of rice. What matters is that the players prefer outcomes with larger numbers attached. We generally refer to a combination of strategy pair and payoff pair in a given game as a potential *outcome*. If the outcome is picked out by a solution concept it is "in the solution set". If the solution concept picks no outcome the solution set is empty.

### Four matrices per game

Games presented this way are often called *matrix games*. The term is imprecise in the sense that a single $2 \times 2$ game may be represented by any of four matrices with the same payoff pairs arranged diagonally opposite. The four matrices can be produced by interchanging rows or columns. The payoff matrices are "identical under an appropriate re-labelling" of the strategies. If players "see through" labels, the games will be behaviourally equivalent. The equivalence may be irrelevant in the real world. If, for example, one strategy is labelled "good" and the other "bad", reversing the labels might well affect behaviour.

The matrix representation illustrates how a formal deep structure can be captured in an apparently simple surface structure. The fact that the rows and columns are at right angles to each other, for example, reflects the idea that the strategies available to the two players are *independent*. Independence means that it is possible to speak of changing Row's strategy without changing Column's strategy. An assumption about the nature of the world is displayed spatially in the matrix.

Other theoretical constructs appear as surface features in the matrix. The *strategy sets* for the two players are shown as the labels for the rows and the columns. The *strategy space*, $S$, is the set of cells in the body of the matrix. Each cell is labelled with its strategy coordinates; $S = \{U,D\} \times \{L,R\} = \{(U,L),(U,R),(D,L),(D,R)\}$. The number of outcomes in the strategy space is simply the product of the numbers of strategies in the strategy sets of the individual players which is the number of cells in the matrix.

The strategy space is the domain of the payoff function. Since the matrix attaches a pair of payoffs to every cell in the strategy space, the bi-matrix is precisely the payoff function, mapping from the strategy space to payoff space. The usual payoff space is the real number plane, $\Re^2$, but for the representative ordinal games it is the discrete space $\{1,2,3,4\} \times \{1,2,3,4\}$.

## 1.4 Summary

The $2 \times 2$ games integrate a remarkable amount of theoretical machinery in a deceptively simple package. This chapter has introduced basic concepts in game theory and key features of the $2 \times 2$ games in strategic form. It has also described a partition of the space of $2 \times 2$ games into a relatively small number of regions, each represented by an ordinal game. The set of ordinal games is the subject of this book, and the task of describing and counting the ordinal games begins in the next chapter.

# Chapter 2

# 144 games

## 2.1 Introduction

In this chapter we introduce a simple way to represent the $2 \times 2$ games. The main innovation is shifting analysis from the space of strategies to the space of payoffs. This dual to the familiar matrix representation provides an alternative introduction to game theory, but it also provides a convenient foundation for the topological analysis of Chapter 3.

We use a device we call the *order graph* and the graphical version of the *inducement correspondence for the Nash situation*. The order graph is an expositional tool that allows us to present and compare games visually. The inducement correspondence is a concept developed by Greenberg [10] that is particularly suited to the payoff-space representation. We combine the two constructs to provide a simple way to count the strictly ordinal $2 \times 2$ games, then go on to explain the way we label individual games.

Numbering systems in the literature are essentially arbitrary, so we introduce a simple indexing system based on the topological relationships between games. A set of four reference figures shows the order graphs for all 144 games arrayed according to our indexing system. We also discuss the patterns that appear at the level of the order graph.

Our main objective in this chapter is to establish terminology and graphical conventions for later chapters. Readers familiar with game theory will find order graphs, the inducement correspondence and our indexing system to be useful tools, and the rest mildly unconventional but elementary. Readers who are less familiar with the $2 \times 2$ games will find the approach provides the framework they need to deal with a huge and constantly expanding literature.

## 2.2 The strategic form in payoff space

The *dual* of the payoff function defined over the strategy space is a representation defined in payoff space. The payoffs to the players are the domain and the inverse function[1] maps payoffs to the strategies that produce them. This dual is particularly useful for our analysis.

| | | *Column's Strategies* | |
|---|---|---|---|
| | | *L* | *R* |
| *Row's* | *U* | 1,4 | 3,3 |
| *Strategies* | *D* | 2,2 | 4,1 |

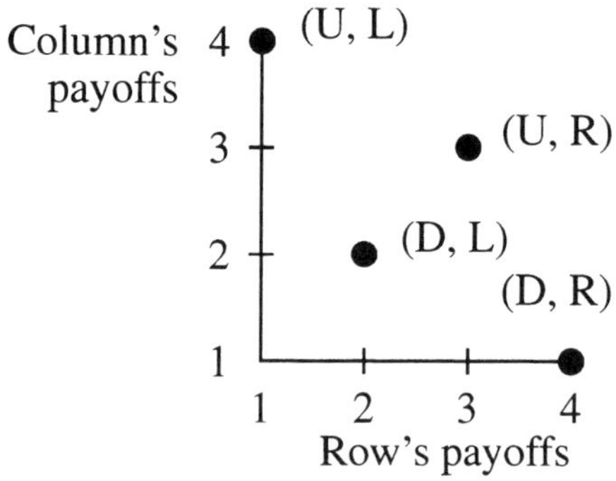


Figure 2.1: Payoff combinations for the Prisoner's Dilemma

A two-person game has a two-dimensional payoff space. In the following development, Row's payoffs are plotted on the horizontal axis and Column's on the vertical axis. Each cell in the payoff matrix represents one possible outcome that appears as a point in the payoff space of the game. Figure 2.1 shows the Prisoner's Dilemma in strategy space and in payoff space.

[1]Since two strategy combinations could yield the same payoffs, we should refer to the inverse *correspondence*, but for the strict ordinal $2 \times 2$ games the problem cannot arise.

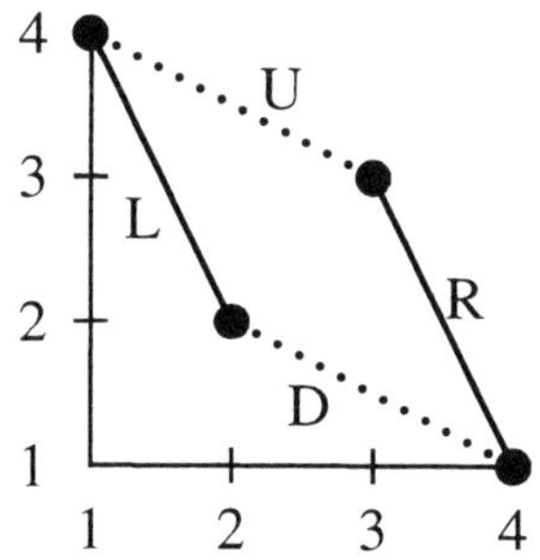


Figure 2.2: Inducement correspondences for the Prisoner's Dilemma

### 2.2.1 The inducement correspondence

Without the labelling the disconnected points in Figure 2.1 do not completely capture the strategic form. The points *UL* and *DL*, for example, are linked in the sense that, by choosing *L*, the column player limits the outcomes available *for the row player* to *UL* or *DL*.

We call the set of payoff vectors when one player's choices are fixed the *inducement correspondence*, short for what Greenberg calls the *inducement correspondence for the Nash situation*. An inducement correspondence is a general term for a set of positions that one player can bring about, or "induce". When we refer to "Row's inducement correspondence" we mean the set of outcomes induced *by* the column player *for* the row player to choose from. Row's inducement correspondences for the Nash situation are always columns of the payoff matrix, and Column's are always rows.

The inducement correspondences for a Prisoner's Dilemma are shown in Figure 2.2 by linking the outcomes in each inducement correspondence with a line. Solid lines identify Row's two inducement correspondences (linking alternatives available to Row once Column has chosen) and dotted lines identify the inducement correspondences for the column player[2].

Identifying the inducement correspondences graphically removes

[2]When hand-sketching these graphs, a dotted line is harder to draw than a solid line, so we use a double line for Row's inducement correspondences and a single line for Column's.

the need to label the points. We now have a complete graphical representation of the the strategic form in payoff space[3]. All outcomes for a $2 \times 2$ game in payoff space are at the intersection of two inducement correspondences, one for each player[4].

Each of the four sides of the quadrilateral in Figure 2.2 corresponds to a row or a column of the payoff matrix. Strategy names can therefore be used to label the inducement correspondences in $2 \times 2$ games, as we have done in Figure 2.2.

Even without the labels, Figure 2.2 is a complete representation of a $2 \times 2$ game in strategic form. Any $2 \times 2$ payoff matrix can be represented as a graph of this sort, and any graph of this sort can be translated into a well-formed payoff matrix. Payoff-space representations for all the ordinal games are presented at the end of the chapter.

### 2.2.2 Using payoff-space representations to analyse games

The inducement correspondence provides a natural unit for analysis and exposition and it yields several of the most fundamental solution concepts in game theory[5].

A *solution set* is simply a subset of the possible outcomes that either predicts how a particular game will turn out or prescribes how it should turn out. A *solution concept* is a rationale for picking a solution based on the information specified in the form. No other information can be used.

---

[3]Figures that plot the payoff vectors are common in the literature, as are figures that show the convex hull of the payoffs. The latter are used for discussing mixed strategies and bargaining games, and require real values. A complete representation of the strategic form requires that the strategic choices represented in the matrix be recoverable, and that is the reason for using different line styles to represent each player's inducement correspondences.

[4]With more than two players, the choices of all but one are fixed. If there are three players, then three inducement correspondences intersect at every outcome in the (three-dimensional) payoff space.

[5]Although we will use the concept of an inducement correspondence throughout the book, we use it in a very limited way. To get a sense of the power of the concept, see Greenberg [10]

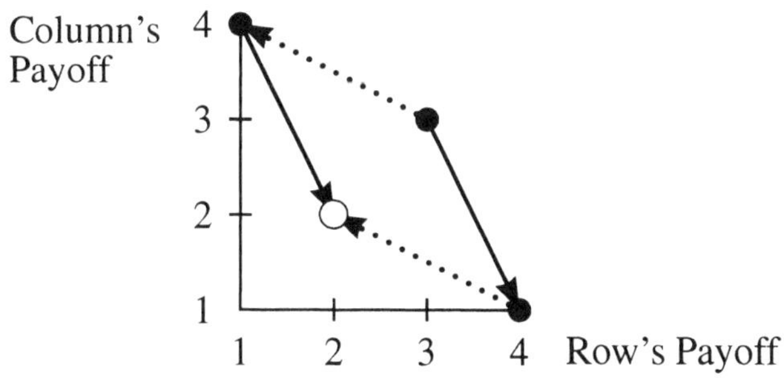


Figure 2.3: Best responses and Nash equilibrium for PD

A solution concept is attractive if

1. it makes sense in its own terms,
2. it seems to provide good advice to players,
3. it yields sensible predictions, or
4. it actually predicts what people will do in situations that correspond to the formal game.

**Nash equilibrium**

The most familiar and most widely accepted solution concept leads to solutions called *Nash equilibria*. From this point on we identify a Nash equilibrium with an open circle as in Figure 2.3.

The Nash equilibrium is often rationalized using a story about how people think and how their behaviour is related to their thoughts. Economists generally assume that, from a set of alternatives, a player will actively choose the one he likes best. This is the assumption of economic rationality, one of the core assumptions of standard game theory. Rationality alone will not predict behaviour in a game, but it leads us to single out the member of any inducement correspondence that yields the greatest payoff for the player that is making the choice. The resulting behaviour is sometimes described as "myopic"[5] because it fails to take into account how other players might respond to a given choice.

## Best response analysis

In the payoff-space representation, this very local version of rationality appears as Row's tendency to select the point in any given inducement correspondence that is farthest to the right. Column selects the *highest* point in an inducement correspondence. We show this in Figure 2.3 by making the lines representing inducement correspondences into arrows pointing toward the preferred points. The arrows indicate the best response for the player choosing in the given inducement correspondence. Solid arrows always point right and dotted arrows point upward. Normally we do not use arrows to identify best responses; the tendency of the players to select the best element in an inducement correspondence is easily read into the diagram.

Inducement correspondences provide a particularly easy way to explore the Nash equilibrium. Because payoffs are strictly ordered there will always be a single best response in a given inducement correspondence. The inducement correspondence can also be used to describe a number of other solution concepts, including maxi-min and solutions based on dominance[6].

Focusing on choice within inducement correspondences is sometimes called *best response analysis*. A player's best response is always her most preferred element in the set made accessible by the actions of other players (ie the set *induced* by them). A *Nash equilibrium* is defined as a best response for all players. A Nash equilibrium is easy to recognize in the payoff space: any payoff pair that is the terminus of two arrows is a Nash equilibrium. In Figure 2.3, the payoff combination for the Nash equilibrium is identified with an open circle.

[6]The maxi-min solution is easy to find. Row will not choose the row with the lowest payoff, so the dotted line that touches the left margin cannot be connected to the maxi-min solution. Column rules out the inducement correspondence that touches the lower boundary of the graph. The maxi-min solution must be therefore at the intersection of the remaining two inducement correspondences. The Periodic Table in Chapter 9 identifies the maxi-min solution for every ordinal $2 \times 2$ game.

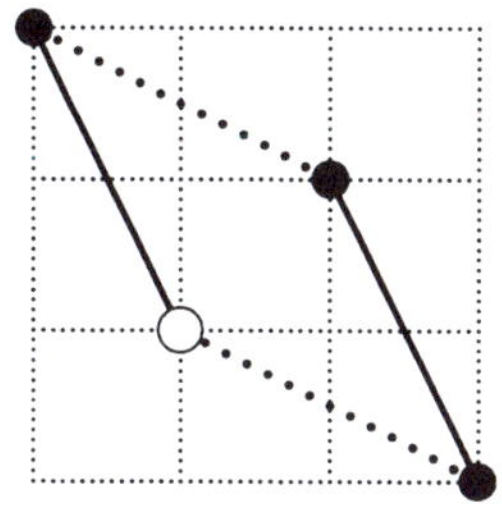

Figure 2.4: Order graph for the Prisoner's Dilemma

## 2.3 Order graphs

We make extensive use of a payoff-space representation designed specifically for ordinal games. The *order graph* for a $2 \times 2$ game is illustrated in Figure 2.4. It is based on the payoff-space representation introduced in Section 2.2. In the payoff-space representation in Figure 2.3, however, whole numbers were used simply because they were convenient. Any point in $\Re^2$ would have been acceptable. In the order graph in Figure 2.4, only the sixteen points on a grid of four vertical and four horizontal lines are acceptable[7]. For any *strict* ordinal game there can be only one point on each vertical and each horizontal line.

### Order graphs and matrices

Every $2 \times 2$ game is represented by an order graph, but each order graph represents an entire equivalence class of games. Furthermore, each order graph can be represented by any one of four matrices. See Figure 2.5. It is convenient to select one of the four as the standard matrix representation of the order graph and the ordinal class of games.

Placing any one payoff pair in a matrix completely determines the positions of the others: selecting the standard matrix involves a single choice. For the graphically inclined, having the matrix ori-

[7] For an $m \times n$ game the grid is $(m \times n) \times (m \times n)$.

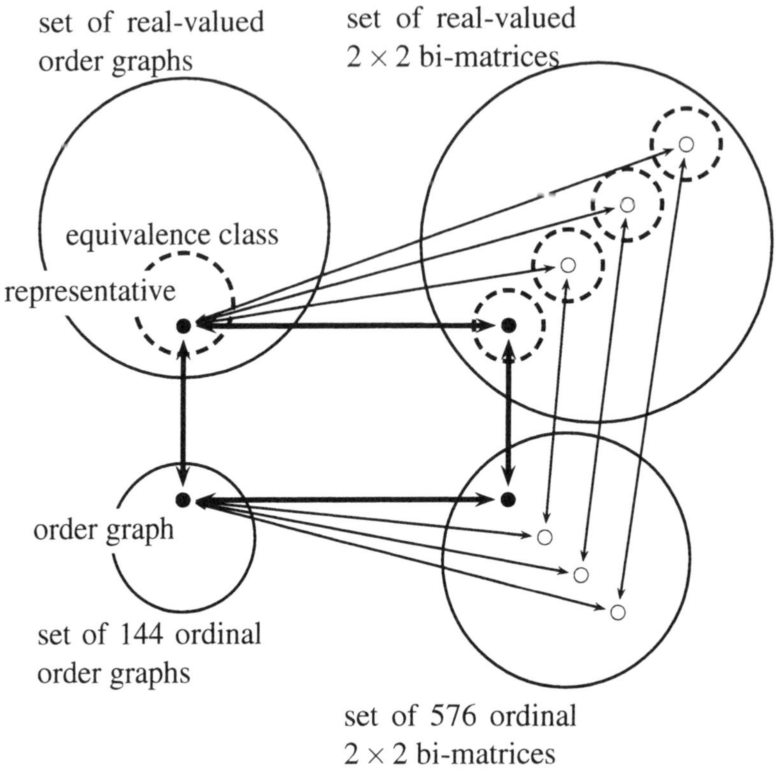


Figure 2.5: Ordinal and real-valued representations of the strategic form in payoff space and strategy space

ented in a manner that is consistent with the graph is desirable, so larger payoffs should be in the upper right cell. There is some ambiguity about what larger means with ordinal pairs, however. The algorithm in Figure 2.6 produces the desired orientation.

## 2.4 Counting the 2 × 2 games

Rapoport and Guyer [23] established a commonly accepted count for the 2 × 2 games. They began with the observation that there are 576 ways to arrange two sequences of four numbers in a bi-matrix. They then decided (in effect) that if matrices produce the same order

**Convention for constructing standard payoff matrices:** Apply the first rule that the payoff matrix allows.

***If the game has a symmetric pair,***

1. (4, 4) ⟶ upper right cell
2. (1, 1) ⟶ lower left cell
3. (3, 3) ⟶ upper right cell
4. (2, 2) ⟶ lower left cell

***If the game has no symmetric payoff pairs, put Row's 4 in the right column AND Column's 4 in the upper row.***

Figure 2.6: Orienting the payoff matrix

graph they are equivalent, reducing the number of games by a factor of four to 144.

They also defined another equivalence they called a "reflection", which amounts to reflecting the order graph around the positive diagonal and reassigning the inducement correspondences. Because their notion of a reflection involves reassigning roles, it is not a conventional geometric or group-theoretic reflection. "Rapoport and Guyer reflections" (R&G reflections) are behaviourally equivalent if players facing the same payoff structure always behave the same way. In other words, reflections are indistinguishable if players are indistinguishable. Imposing these two equivalencies, they counted 78 distinct games.

### 2.4.1 Using order graphs to count the 2 × 2 games

The inducement correspondence in payoff space provides an alternative and possibly more intuitive approach to counting the 2 × 2 games.

We can begin with any payoff pair. It shares an inducement correspondence with another payoff pair. That payoff can be at any

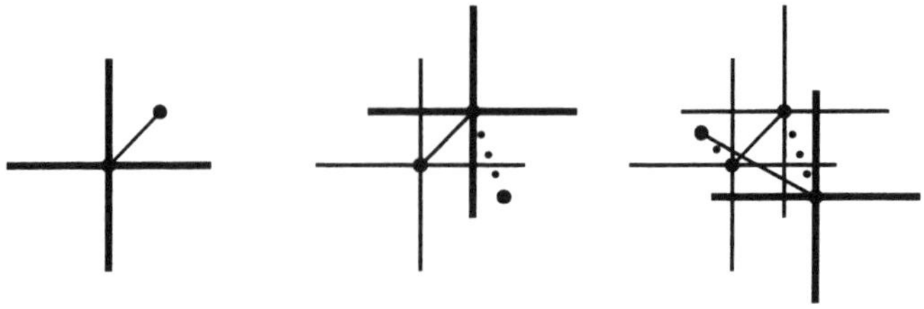

Figure 2.7: An approach to counting the $2 \times 2$ games

distance in any direction from the first. With strict orderings, the second payoff pair must be in the interior of one of the four quadrants formed in Figure 2.7(a) by drawing vertical and horizontal lines through the original point. In Figure 2.7(a) we have arbitrarily selected a point in the upper right quadrant.

We have now chosen two of the four payoff vectors required to define a $2 \times 2$ game. Without loss of generality we can arbitrarily assign the two points to Row's inducement correspondence (one of the columns of a payoff matrix).

Figure 2.7(b) shows that there are 9 alternatives for the location of the third point, since it can be below the lowest, between the two previous points, or above both of them, and it can also be either left of, right of, or between them. The second and third point form an inducement correspondence for the column player. They represent one of the rows of a payoff matrix. We can imagine that the first point represented the lower left cell of the the payoff matrix. The second point would then represent the upper left cell, and the third would be in the upper right.

By the same reasoning, there are 16 ways to choose the fourth point (c), which represents the lower right cell in the payoff matrix. The number of games is therefore $4 \times 9 \times 16 = 576$. A simple relationship has emerged – the number of games that we have to consider is $((2 \times 2)!)^2$. This approach can be applied to games of any size[8].

Some of these games are duplicates. It is possible to produce

[8] In the $2 \times 3$ game we choose six points, and there are $((2 \times 3)!)^2$ such games.

exactly the same figure with the same player owning the first inducement correspondence by starting from any of the four points. There are, therefore, only $576 \div 4 = 144$ distinct games.

The 144 games might be reduced further, following Rapoport and Guyer[23], by eliminating R&G reflections. There are strong arguments against eliminating reflections that we will take up in Chapters 3 and 9, but the convention has been that there are only 78 distinct $2 \times 2$ games[9].

### 2.4.2 Numbering the 2×2 games

Numbering systems in the literature are completely arbitrary so we have developed an indexing system that reflects the topological structure of the $2 \times 2$ games. Each game's number provides information about the game and can be used to find related games quickly. The indexing system serves as a first, rough map of the topological space introduced in the next chapter.

Each game has a three-digit index. The Prisoner's Dilemma, for example, is $g_{111}$. Each digit in the subscript corresponds to one of three features of the payoff matrix:

| INDEX | FEATURE | NUMBER |
|---|---|---|
| c | the column player's payoff pattern | 6 |
| r | the row player's payoff pattern | 6 |
| $\ell$ | the relative orientation of the two payoff patterns | 4 |

We call $c \in \{1,2,3,4,5,6\}$ and $r \in \{1,2,3,4,5,6\}$ the column and row indices. For reasons that will become clear, we call $\ell \in \{1,2,3,4\}$ the *layer* index. Symmetry between rows and columns leads us to begin indexing with a symmetric game. We have chosen to count

[9] To reproduce Rapoport and Guyer's count we need to eliminate "reflections". Notice first that any game with an order graph that is symmetric about the positive diagonal is its own reflection, and second, that there can be only 12 such games. A symmetric game must have two payoffs on the positive diagonal {(1, 1), (2, 2), (3, 3), (4, 4)}. There are ${}_4C_2 = 6$ ways to achieve this. Having chosen two symmetric points there are only two symmetric ways to join them to the remaining two points. $2 \times {}_4C_2 = 12$. Therefore there are $144 - 12 = 132$ asymmetric games. Every asymmetric game has a reflection. There are therefore $(132 : 2 + 12) = 78$ distinct games if reflections are equivalent.

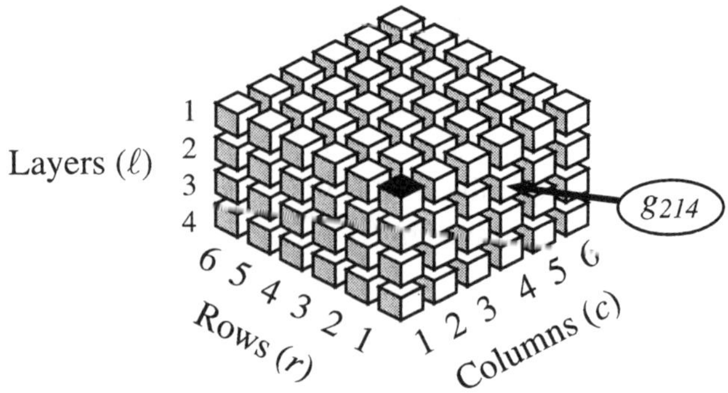


Figure 2.8: Arrangement of the $2 \times 2$ games by indices, $g_{\ell rc}$

rows, columns and layers beginning with the (symmetric) Prisoner's Dilemma because it is the best known game of all.

With six column patterns, six row patterns, and four layers, we have room for exactly 144 $2 \times 2$ games. The entire collection of games can be visualized, as in Figure 2.8, as an array that is six games wide, six games high and four layers deep. The three-digit index locates each game in the array.

## 2.5 All 144 games

Figures 2.13 to 2.16 show the order graphs for all 144 games. Figure 2.14 contains the 36 games in layer 1. Figures 2.13, 2.15 and 2.16 show layers 2, 3 and 4. Even without the topological relationships that we introduce in the next chapter, these figures provide a convenient presentation of the $2 \times 2$ games.

The appendix to this chapter describes the six patterns and the differences between layers more fully. Table 2.2 in the appendix will be a useful reference for the following chapter, but it provides more detail than we need to proceed.

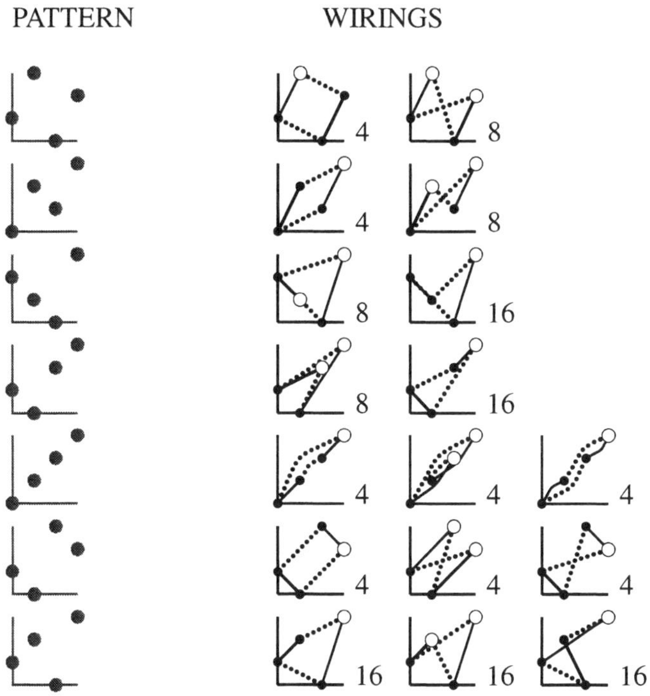


Figure 2.9: Elementary payoff patterns, wirings, and numbers of isometries

### 2.5.1 Types of order graphs

Order graphs are patterns drawn on a four-by-four grid. They consist of four nodes connected by two pairs of lines to form a quadrilateral. Each of the nodes represents a payoff pair; each line represents a row or a column of a payoff matrix. Strict ordinality implies that no vertical or horizontal line in the grid can contain more than one dot.

Figure 2.9 shows the seven basic payoff patterns from which all 144 games can be constructed. To construct a particular order graph we select the appropriate elementary patterns, connect the dots, assign the inducement correspondences, and then rotate or reflect the resulting quadrilateral to get the right orientation. Rotations and reflections preserve the shape of the figure but change its orientation. Any point-to-point transformation that preserves distances and relations between points in this way is called an *isometry*.

| Symbol | Description of transformation |
|---|---|
| $I$ | orientation unchanged |
| $R^{\Uparrow}$ | rotation 90° counterclockwise |
| $R^{\Updownarrow}$ | rotation 180° |
| $R^{\Downarrow}$ | rotation 90° clockwise |
| $R^{\nwarrow}$ | reflection across positive diagonal |
| $R^{\swarrow}$ | reflection across negative diagonal |
| $R^{\leftarrow}$ | reflection across vertical centreline |
| $R^{\downarrow}$ | reflection across horizontal centreline |

Table 2.1: Transformations that map the order graph grid onto itself

To understand the patterns in the set of 144 order graphs, we need to understand how the dots can be wired, how they can be assigned, and the various ways that the figures can be rotated or reflected.

### Wirings

For some patterns, there are three distinct "wirings"; for others, only two. Figure 2.9 shows the seventeen distinct quadrilaterals and the number of times each appears among the 144 games.

For example, the first row shows the two quadrilaterals constructed with the points $(1,2)$, $(2,4)$, $(3,1)$, $(4,3)$. The first looks like a square tilted right, the second like an hourglass.

The number 4 beside the square indicates that four games can be made from it by reflection, rotation or reassignment. Eight games can be made from the hourglass figure.

### Assignment

If the dotted lines in the square in the first row of Figure 2.9 are changed to solid lines, and the solid lines to dotted lines, the result is a new game. In the new game Row faces the alternatives that Column faced in the original game. We call the new game the *antigame* of the original, and we get to the new game by the operation $A$, which reassigns the inducement correspondence.

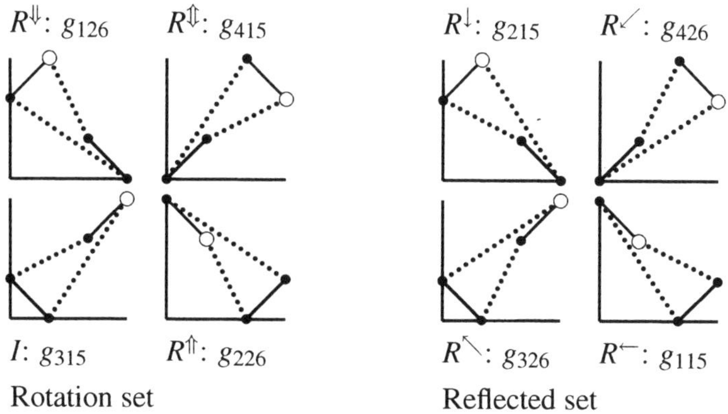


Figure 2.10: Transformations of $g_{315}$

Unlike the geometric transformations, $A$ is not isometric because distances between points are not preserved. In the hourglass figure in the first row of Figure 2.9, transformation $A$ maps one side of the square to the diagonal, which is longer. In the payoff bi-matrix, exchange of inducement correspondences is achieved by exchanging payoff vectors between a pair of diagonally opposite cells.

### Rotations and reflections

Rotations and reflections are not inherently interesting for game theorists, but they are associated with regularities in the space of the $2 \times 2$ games. They are in fact the key to understanding the symmetries exhibited by the $2 \times 2$ games.

Imagine first the four-by-four grid without a game drawn on it. If it is rotated $90^{\circ}$ clockwise the resulting figure is indistinguishable from the original. The grid is also mapped onto itself by reflections around vertical, horizontal, or diagonal lines through its centre. In all, three rotations and four reflections leave the grid looking the same. When a figure is unchanged by an isometry we say it exhibits a *symmetry*. The transformations are listed in Table 2.1 along with the identity operation. There are no other orientations that look the same as the original grid.

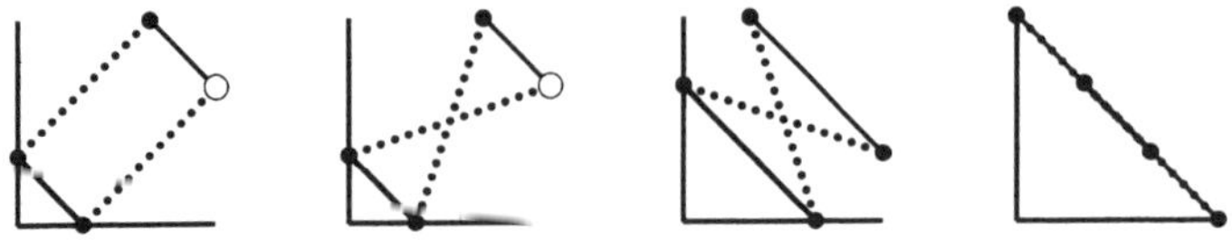


Figure 2.11: Some quasi-symmetric games

Now consider reflections and rotations of the square in the first row of Figure 2.9. Rotating the figure by 90° produces the same effect as reassignment. Rotation by 180° returns us to the original figure. Rotation by 270° adds no new possibilities.

Reflection in the vertical line, $R^{\leftarrow}$, produces a distinct quadrilateral, a square tilted slightly left. A diagonal reflection produces a reassignment of the same shape. From all the transformations there are only four distinct variants based on the square wiring.

The second quadrilateral in the first row can be oriented four ways ($I$, $R^{\Uparrow}$, $R^{\leftarrow}$, $R^{\nwarrow}$) so eight games are represented. No third quadrilateral is possible.

The completely asymmetric quadrilaterals in the bottom row of Figure 2.9 are distinguishable under all transformations and therefore appear in $8 \times 2 = 16$ games[10]. Figure 2.10 shows a set of order graphs based on a strictly asymmetric game. Eight more are generated by exchanging inducement correspondences.

### 2.5.2 Quasi-symmetric games

The only transformation that has any obvious economic interpretation is $R^{\nwarrow}$, reflection in the positive diagonal. This transformation exchanges payoffs between players in each outcome. In strategy space, $R^{\nwarrow}$ is manifested as a reordering of payoffs in each cell of the bi-matrix. Row comes to care only for Column's payoffs. Perhaps it's love.

[10]In fact, the pattern on the bottom row accounts for one third of all games with three asymmetric quadrilaterals.

To see how this reflection relates to the overall organization of the space of the $2 \times 2$ games, notice that every order graph in the upper right of each layer in Figures 2.13 to 2.16 has an $R^{\nwarrow}$ reflection in the lower left of the same layer. Each 36-game layer can be folded along its negative diagonal so that the game in the upper right coincides with its $R^{\nwarrow}$ reflection in the lower left.

The 24 games on the negative diagonals of the layers are invariant under $R^{\nwarrow}$. Some examples are shown in Figure 2.11. Although $R^{\nwarrow}$ does not change the order graph, and the graphs are obviously symmetric, these games are not symmetric in the game-theoretic sense. (See page 58.) We therefore call them *quasi-symmetric.*

### 2.5.3 Assignment and reflection

Figure 2.12 shows the effect on game $g_{315}$ of $A$, of $R^{\nwarrow}$, and of applying both transformations in sequence. The result does not depend on order of application:

$$R^{\nwarrow}(A(g_{315})) = A(R^{\nwarrow}(g_{315})) = g_{351}$$

The combined transformation maps $g_{315}$ into its R&G reflection. Earlier authors have defined games related by this compound transformation as identical.

The 12 symmetric games are the only games that are invariant under the compound transformation. For these games, the two transformations are equivalent so one "undoes" the other. For example (Prisoner's Dilemma)

$$R^{\nwarrow}(A(g_{111})) = g_{111}$$

## 2.6 Summing up

In this chapter we have introduced an approach based on representing the games in payoff space. Order graphs allow us to describe games easily, and our indexing system lets us lay out the games in a systematic and revealing way. Symmetries in the order graph representation shed some light on the nature of symmetric and reflected games, and on the structure of the space of the $2 \times 2$ games. They are not directly useful for analysing behaviour.

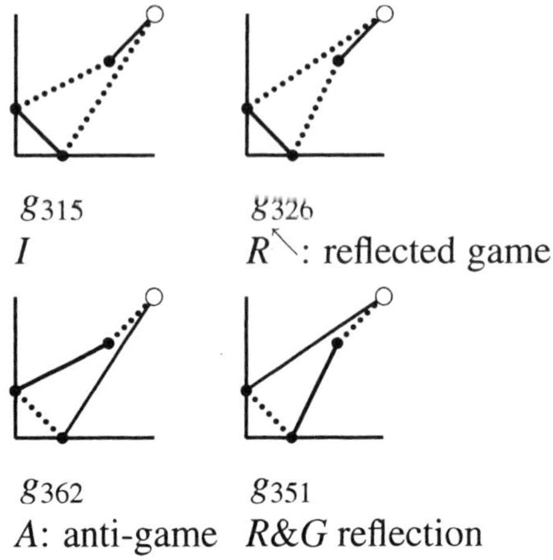


Figure 2.12: Reflections ($R^{\nwarrow}$) and reassignments ($A$)

We are now ready to begin the real work of the book, which is to explore our alternative approach to the relationships among the 144 representative games.

# 2.7 Appendix: Relating payoff patterns to the indexing system

### Row, column and stack

If we arbitrarily select the location of one player's most preferred outcome in the payoff matrix, we can produce six distinct payoff patterns for that player by permuting the positions of the three less desirable payoffs. The bottom row of payoff matrices in Table 2.2 illustrates the six basic patterns for the column player, keeping the 4 payoff in the upper left cell. Asterisks indicate that the row player's payoffs are the same as in the first game in the row (are invariant in the row). The first game in this example is the Prisoner's Dilemma.

**A column**

| $g_{161}$ | |
|---|---|
| 1,* | 2,* |
| 3,* | **4**,* |

| $g_{151}$ | |
|---|---|
| 2,* | 1,* |
| 3,* | **4**,* |

| $g_{141}$ | |
|---|---|
| 3,* | 1,* |
| 2,* | **4**,* |

| $g_{131}$ | |
|---|---|
| 3,* | 2,* |
| 1,* | **4**,* |

| $g_{121}$ | |
|---|---|
| 2,* | 3,* |
| 1,* | **4**,* |

**A stack**

| | $\ell = 2$ | $\ell = 3$ | |
|---|---|---|---|
| column | $g_{211}$<br>1,2 3,1<br>2,**4** **4**,3 | $g_{311}$<br>1,1 3,2<br>2,3 **4**,**4** | both |
| | $g_{111}$<br>1,**4** 3,3<br>2,2 **4**,1 | $g_{411}$<br>1,3 3,**4**<br>2,1 **4**,2 | row |
| | $\ell = 1$ | $\ell = 4$ | |

| $g_{111}$ | | $g_{112}$ | | $g_{113}$ | | $g_{114}$ | | $g_{115}$ | | $g_{116}$ | |
|---|---|---|---|---|---|---|---|---|---|---|---|
| *L* | *R* | | | | | | | | | | |
| 1,**4** | 3,3 | *,**4** | *,3 | *,**4** | *,2 | *,**4** | *,1 | *,**4** | *,1 | *,**4** | *,2 |
| 2,2 | **4**,1 | *,1 | *,2 | *,1 | *,3 | *,2 | *,3 | *,3 | *,2 | *,3 | *,1 |

**A row**

Table 2.2: Payoff patterns in the $2 \times 2$ games

The second matrix in the row was produced by swapping the positions of the 1 and the 2 in matrix 1. The third is produced by swapping the 2 and the 3 in matrix 2. The fourth is produced by swapping the 1 and the 2 in matrix 3. For the fifth, swap 2 and the 3 in matrix 4, and for the sixth swap 1 and the 2 in matrix 5. We explain the reason for the particular order of permutations in the next chapter.

The same procedure can be applied to the row player's payoffs. The results are shown in the matrices in the left-hand column of Table 2.2.

Thirty-six combinations can be constructed with the six row and six column patterns. Each of these 36 games has the most preferred element for each player in exactly the same position in the payoff matrix. For the example in Table 2.2 the most preferred elements are all diagonally opposite.

Now consider alternative ways of locating the column player's highest payoffs. Table 2.3 shows the effect of column or row operations on the payoff matrix for the column player. These exchanges maintain the sequence of payoffs around the payoff matrix, changing only the starting point and/or the direction. The 4 and the 1 remain diagonally opposite.

The changes shown in Table 2.3 can be described in an interesting way as the result of two reflections. A reflection is an isomorphism that conserves relative positions of elements but reverses their positions relative to a line. Pattern 2 is the result of reflecting the col-

1) Pattern 1

| *L* | *R* |
|---|---|
| $*,4$ | $*,3$ |
| $*,2$ | $*,1$ |

4) Rows exchanged

| *L* | *R* |
|---|---|
| $*,2$ | $*,1$ |
| $*,4$ | $*,3$ |

2) Columns exchanged

| *L* | *R* |
|---|---|
| $*,3$ | $*,4$ |
| $*,1$ | $*,2$ |

3) Rows, columns exchanged

| *L* | *R* |
|---|---|
| $*,1$ | $*,2$ |
| $*,3$ | $*,4$ |

Table 2.3: Patterns equivalent to column payoff pattern 1

umn player's payoff matrix around a line running between the rows. Pattern 4 is a reflection around a line running between the columns. Pattern three is a combination of both reflections, which results in a 180° rotation.

Combining any *row* pattern with the four variations on column pattern produces four *different* games. The payoff matrices for the four games based on the Prisoner's Dilemma are shown in the upper right of Table 2.2.

What if the pattern of the row player's payoffs were manipulated instead of the column payoffs? Any games produced would be equivalent to one of the four already produced. Row pattern one and column pattern one can be combined to create only *four* distinct games. We call the set of games generated from one pair of payoff patterns a *stack*. Games in a stack share row and column indices, but each has a distinct *layer* index.

Table 2.2 is a useful reference for Figures 2.13 to 2.16 and Figures 3.4 in the following chapter because it provides a view in terms of payoff matrices. In the following chapters arguments will generally be presented in terms of operations on order graphs.

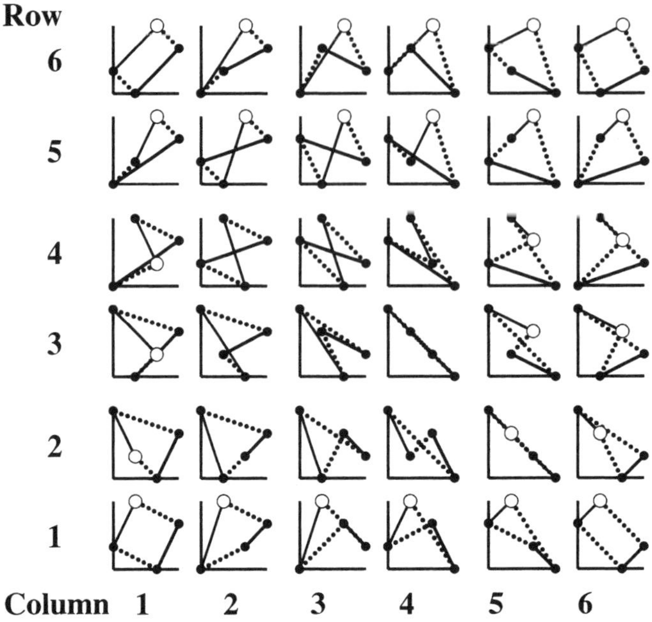


Figure 2.13: **Layer 2**: Column generally does well

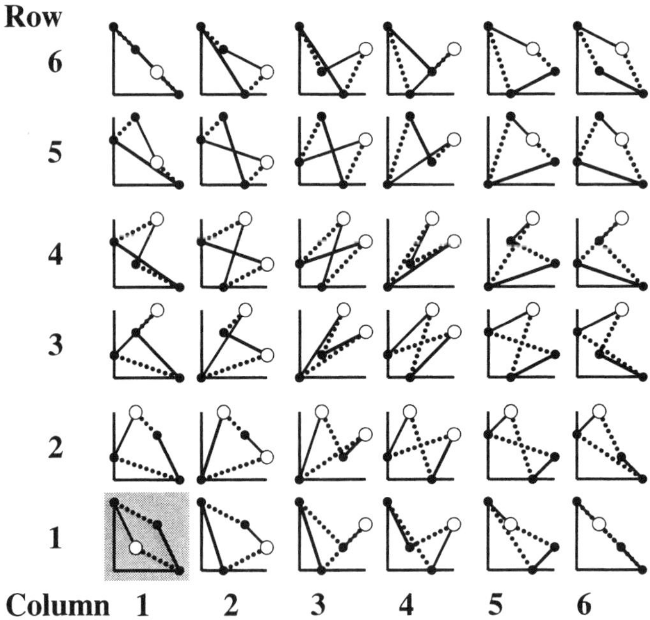


Figure 2.14: **Layer 1**: With PD, Chicken and BoS
Game index: $g_{layer, row, column}$

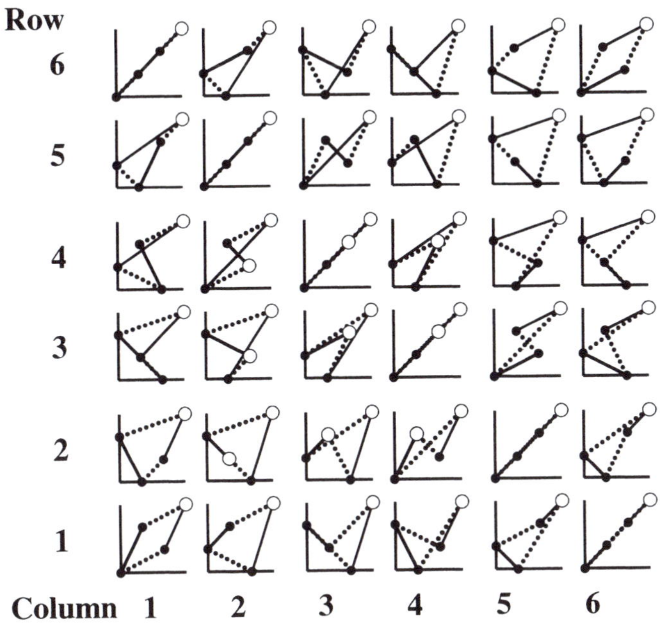


Figure 2.15: **Layer 3**: Best outcome feasible for both

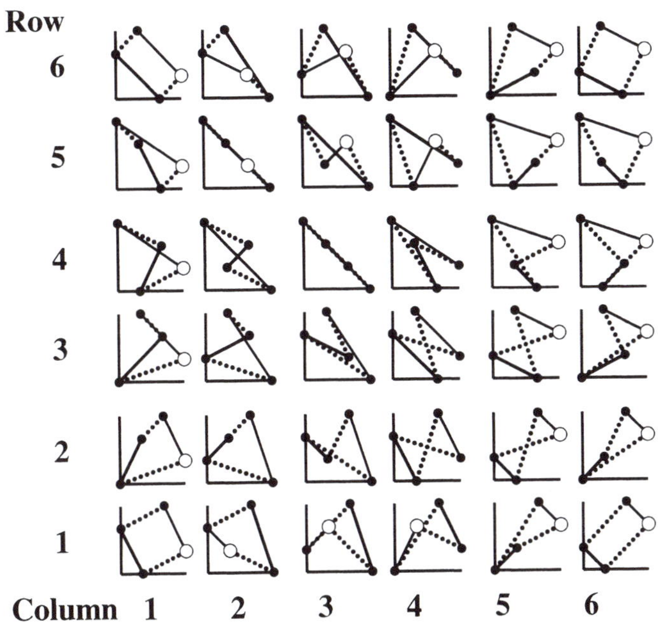


Figure 2.16: **Layer 4**: Row generally does well

# Chapter 3

# Elementary topology of $2 \times 2$ games

> In every subject one looks for the topological and algebraic structures involved, since these structures form a unifying core for the most varied branches of mathematics.
>
> Weise and Noack, "Aspects of topology" ([41] p. 593)

In Chapter 2 we described the $2 \times 2$ games in terms of the payoff function. Payoff functions provide a complete description of a game in strategic form. We now introduce enough additional structure to induce a topology on the set of games as a whole. The topology allows us to relate the 144 $2 \times 2$ games in a new and systematic way.

Every game is related to every other in the sense that there is a transformation that converts the payoff structure for one into the payoff structure for the other. A complete graph, showing all transformations among the 144 ordinal games as $10,296$ edges would be easy to create but essentially useless. What we want is a graph that shows similar games near each other and different ones widely separated. We begin by restricting the set of transformations to six; that is, there are six transformations whose rules can be applied to any game to produce six adjacent games.

The resulting graph has only 432 edges. It is still possible to start with any game and reach any other via some sequence of transformations, but some games, called *nearest neighbours*, can be reached via a single transformation while others require several steps. Because the graph is connected, the six transformations constitute a *set of generators*[1].

When the graph is embedded within a surface it is called a *map*. The nature of the surface needed to embed the graph without crossing edges is a topological feature, and the topological structure of this payoff *space* is not only useful, but also beautiful.

Our goal in this chapter is to develop and explain the graph, which serves as a blueprint for the space of ordinal $2 \times 2$ games. In section 3.2 we develop the appropriate concept of a neighbourhood for the $2 \times 2$ games and examine the neighbourhood of a specific game. In section 3.4 we build up a picture of the graph of the $2 \times 2$ games beginning with the simplest subspaces and gradually combining all six transformations. At several points we describe the topology of special subspaces and their associated subgraphs.

In many cases the simplest language for describing the subspaces and the subgraphs comes from group theory, so we introduce elementary terms and concepts where they are useful. The terms are explained as they occur and brief definitions are provided in the Glossary. No advanced mathematics is required.

Two features of the topology deserve special attention. First, the definition of the six transformations is rooted in the structure of *preferences*. Second, economically related games are *near* each other in the graph. In fact, they often occupy well defined subspaces. These subspaces are generated by subsets of the six transformation.

Chapters 4 to 8 will examine the games in specific subspaces in detail and Chapter 9 will deal with the most general features of the topological space of the $2 \times 2$ games.

---

[1]It is not, however, a minimal set of generators. Subsets of the transformations can be used and the graph will still be connected, but economically significant links will be lost.

## 3.1 About topologies

Topology is the mathematical study of properties of objects which are preserved through deformations, twistings, and stretchings but not through breaks or cuts. Most introductions to topology begin by considering deformations of Euclidian surfaces like the torus or the Möbius strip. This approach treats topology in terms of "one-to-one bi-continuous mappings of sets of points in Euclidian space" [41]. It is an unnecessarily special approach. One can instead generate a topological space from an arbitrary set of abstract elements, called *points*, by imposing a topology on the set [18]. In our case the set of points is the set of $2 \times 2$ games.

Any set for which a topology has been specified is called a topological space ([18], p. 76). Modern topology in fact takes no account of the individual nature of the elements, but merely of their mutual relationships. In an approach based on points, however, what is meant by the expression *neighbourhood of a point* must be defined axiomatically. In his *Grundzüge der Mengenlehre*, Hausdorff [12] defined his concept of a topological space. The four Hausdorff axioms are:

1. To each point $x$ there corresponds at least one neighbourhood $U(x)$, and $U(x)$ contains $x$.
2. If $U(x)$ and $V(x)$ are neighbourhoods of the same point $x$, then there exists a neighbourhood $W(x)$ of $x$ such that $W(x)$ is a subset of the union of $U(x)$ and $V(x)$.
3. If $y$ is a point in $U(x)$, then there exists a neighbourhood $V(y)$ of $y$ such that $V(y)$ is a subset of $U(x)$.
4. For distinct points $x$ and $y$, there exist two disjoint neighbourhoods $U(x)$ and $V(y)$ [42].

Developing a satisfactory notion of *neighbour* is the key to developing a topological treatment of the $2 \times 2$ games. Preferences as we think of them in economics provide enough structure to induce a topology on the ordinal $2 \times 2$ games.

$g_{111}$

| L | R |
|---|---|
| **1**,4 | 3,3 |
| **2**,2 | 4,1 |

⇓

| L | R |
|---|---|
| **2**,4 | 3,3 |
| **1**,2 | 4,1 |

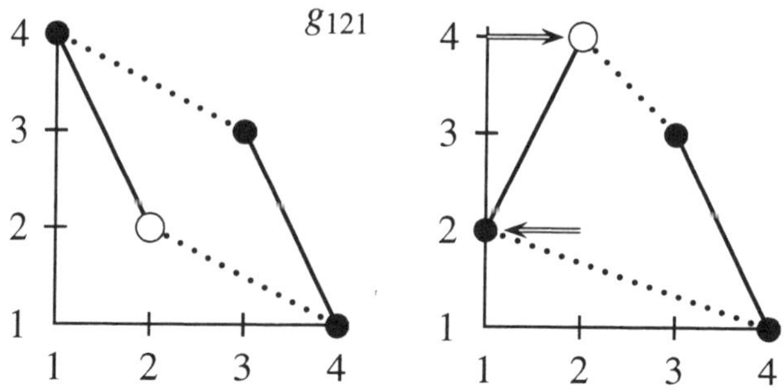


Figure 3.1: Changing the ordinal values for the two lowest-ranked outcomes for the row player

## 3.2 What is a neighbour?

Since games are characterized by the payoff function, similar games must have similar payoff functions. To define meaningful neighbourhoods, we need to characterize the smallest significant change in the payoff function. Obviously a change affecting the payoffs of one player is smaller than a change affecting two players. The closest neighbouring games are therefore those games that differ only by a small change in the ordering of the outcomes for one player.

At this point the structure of preferences becomes relevant. The outcome liked least by a player has a rank 1. If that outcome becomes more and more attractive it will eventually be preferred to the outcome with rank 2. When this switch occurs, the effect on the payoff matrix is to exchange the positions of the 1 and 2. This is an example of a minimal transformation yielding a different, neighbouring game.

The set, $\boldsymbol{S}$, of minimal exchanges has six members:

$$\begin{aligned} \boldsymbol{S} &= \{X_{ij} | i \in \{1..3\}, j = i+1, X \in \{R, C\}\} \\ &= \{R_{12}, R_{23}, R_{34}, C_{12}, C_{23}, C_{34}\} \end{aligned}$$

where $X_{ij}$ changes the rank of the outcome originally ranked $i$ by the player $X$ to rank $j$ and the rank of the outcome originally ranked $j$ to rank $i$. When $X = R$, we call it a *row swap* indicating Row's preferences have changed and if $X = C$ it is a *column swap*.

Figure 3.1 illustrates $R_{12}$, a '1 for 2' swap for the row player in the Prisoner's Dilemma. Ranks that change are shown in boldface[2]. Notice that swaps are applied to the ordinal payoffs wherever they appear in the payoff matrix. The effect of the change in the payoffs is illustrated in the order graphs on the right.

Changes like this in the payoffs might result from *small* changes in information, preferences, or technology, or from small errors in identifying games. A player might, for example, receive a very small amount of new information. She might then reconsider the outcome she had originally ranked 1, and decide that it is slightly better than she first thought, and that it is superior to the outcome she had previously ranked 2. She would naturally relabel the two outcomes, resulting in a different payoff matrix, and hence a different game. The new game is close to the old game in that it is reached as a result of a small perturbation in one player's information set. The game is also close in the sense that it might be mistaken for the original game or it might evolve into the other as a result of a small exogenous change in the underlying technical conditions.

Examining the neighbours of a game can also provide evidence on the robustness of solutions in the face of perturbations in the payoff structure. In Figure 3.1 the smallest possible change in Row's perception of the least-liked outcomes leads to a change in his behaviour, transforming the PD, with its inefficient equilibrium, into a game with a Pareto-efficient equilibrium.

Since any of the three swaps can be applied to payoffs for either player, it follows that every game has exactly six nearest neighbours. Preferences imply a structure of overlapping neighbourhoods and induce a topology on the set of games. Games can be characterized as close or distant neighbours depending on how many swaps are necessary to transform one into the other.

### 3.2.1 Talking about the neighbours

A swap is a mapping from the space of $2 \times 2$ games to itself: $X_{ij}(g) = h$ , where $g$ and $h \in G_{144}$, the set of games. The primary *neigh-*

[2]The new game has been described and named several times. (See Table 9.1). Brams [5] uses it to describe the Polish Crisis, 1980 1, and the Union – Confederacy crisis.

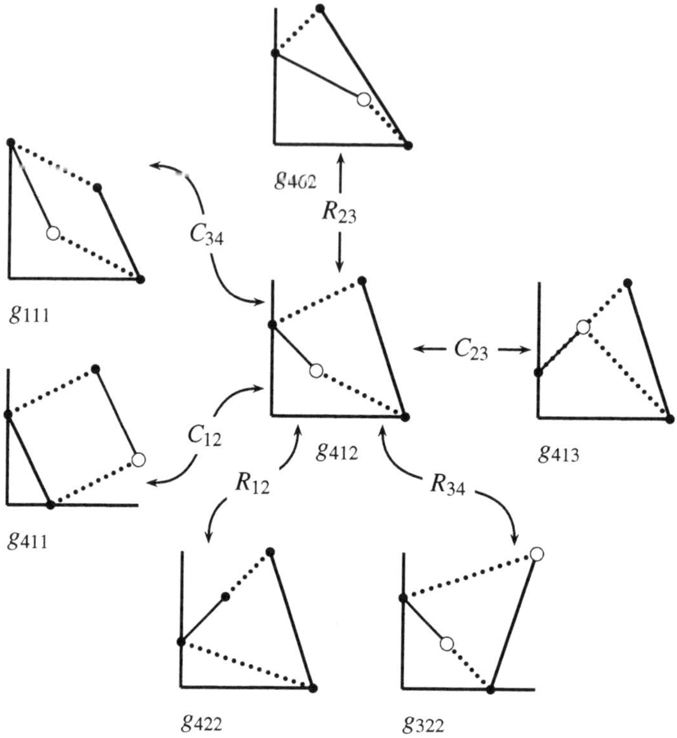


Figure 3.2: The neighbourhood of $g_{412}$

*bourhood* of a game is the set of games that can be reached by a single swap:

$$N_1(g) = \{X_{ij}(g) \mid X_{ij} \in \mathbf{S}\}, g \in G_{144}$$

Neighbourhoods are thus defined strictly in terms of the preferences of the players.

Figure 3.2 shows the immediate neighbourhood of $g_{412}$, a game of interest in its own right. Like the Prisoner's Dilemma, it has a single Pareto-dominated Nash equilibrium but unlike the Prisoner's Dilemma, only one player has a dominant strategy. The game is clearly closely related to the Prisoner's Dilemma in important ways. It will generate social dilemmas as interesting as those generated by the Prisoner's Dilemma.

### A bad neigbourhood?

The neighbourhood $N_1(g_{412})$ contains the Prisoner's Dilemma ($g_{111}$) and $g_{413}$. All three games have Pareto-dominated unique Nash equilibria. We can conclude that

> There is a connected region containing games like the Prisoner's Dilemma with nasty outcomes.

This can be seen as the first result derived by construction using the topological approach. We will show in Chapter 5 that the region contains seven games, only one of which, the Prisoner's Dilemma, is symmetric.

### Strange neighbours

Row swaps from $g_{412}$ yield additional insights. One neighbour, $R_{12}(g_{412}) = g_{422}$ has no Nash equilibrium in pure strategies[3]. Another neighbour, $R_{34}(g_{412}) = g_{322}$, has two Nash equilibria. It belongs to another group, as we show below, yielding a surprising fact:

> A game with no pure strategy equilibrium can be two minimal steps from a game with two equilibria.

The significant observation is that for some games the equilibrium can be quite fragile. The topological approach provides a way to examine the robustness of payoffs and strategies by seeing how they vary for neighbouring games. Of the six perturbations of $g_{412}$ that form $N_1(g_{412})$, four leave the equilibrium strategy combination unchanged and two leave the Nash equilibrium payoffs unchanged. In one of new games, however, a new and Pareto-superior Nash equilibrium emerges, so payoffs are likely to change even though the original payoffs are still a Nash equilibrium. In $g_{422}$ the Nash equilibrium disappears. Only one of the six swaps leaves the payoffs unchanged, so equilibrium payoffs are *not* a robust feature of $N_1(g_{412})$. Equilibrium strategies are reasonably robust in the neigbourhood.

---

[3]Chapter 9 will show that the games with no Nash equilibria also form a connected set. Furthermore, the Prisoner's Dilemma family of games with inferior equilibria lies on the boundary of this important set of games.

### Symmetric neighbours

Another fact emerging from an examination of this small neighbourhood is that $g_{412}$ has two *symmetric games* as neighbours ($g_{322}$ and $g_{111}$, the Prisoner's Dilemma). It is possible to make one symmetric game, the Prisoner's Dilemma, into another symmetric game by the sequence of operations $R_{34}$, $C_{34}$. We call a combined swap operation in which the same swap is made for the row and the column players, a *symmetric operation*, $S_{ij}$. In the example here,

$$S_{34}(g_{111}) = C_{34}(R_{34}(g_{g_{111}})) = R_{34}(C_{34}(g_{111})) = g_{322}$$

Note that a symmetric operation preserves symmetry if symmetry is present initially.

Under symmetric operations, the symmetric games form a subspace which we explore in Chapter 4. In addition, symmetric operations beginning with non-symmetric games generate other subspaces of interest.

## 3.3 Groups

It is sometimes revealing to describe certain sets of games and the corresponding subgraphs in terms of mathematical groups. We therefore introduce some basic terminology from group theory before proceeding to develop the graph of the 2 × 2 games. A *group*, $G$, is a set of elements and a binary operation which together satisfy the four fundamental properties of closure, associativity, the identity property, and the inverse property. Any single swap operator, $X_{ij}$, plus an identity operation, say $I$, can be the elements of a simple group.

Discrete clock arithmetic is a convenient example of a group. There are 12 hours on the face of a clock. The positions on the clock are not the elements of the group, however. The elements of the group are better understood as rotary advances of one hour, two hours and so on. The element 3, for example, represents moving a quarter turn in the clockwise direction from any position on the clock. Any element can be combined with any other element: $4+1=5$, or $5+1=6$.

1. **Closure:** $\forall A, B \in G : AB \in G$.
2. **Associativity:** $\forall A, B, C \in G : A(BC) = (AB)C$.
3. **Identity:** $\exists I \in G, \forall A \in G : IA = AI = A$.
4. **Inverse:** $\forall A \in G, \exists B \in G : AB = I$.

Table 3.1: Properties of a group

*Closure* requires that combining any two elements, such as 7 and 8 yield another member of the group, which is 3 in this case. We can easily write down a table of addition showing, for example, that $6 + 11 = 5$, which ensures that the group is closed under the operation. *Associativity* requires that $(1+2)+3 = 1+(2+3)$.

A group must have an *identity* element that satisfies $I + x = x$ for any member $x$ of the group. An obvious identity element for clock arithmetic is 0 hours[4]. The *inverse* of an element $x$ is an element $y$ that, when added after $x$, returns us to our original position. For the 12-hour clock the inverse of $x$ is $12 - x$. The inverse of 1 is 11, since $1 + 11 = 0$, and the inverse of 6 is 6.

The group can be written $G_{12} \equiv \{0,1,2,3,4,5,6,7,8,9,10,11\}$ under the operation $+$. The subscript 12 is called the *order* of the group and is simply the number of elements in the group. The identity element is 0.

The concept of a *set of generators* for a group is especially useful. A set of generators is a subset of the group that can be used, by repeated application, to generate all the members. A group may have more than one set of generators.

One set of generators for $G_{12}$ is the set $\{0,1\}$. The set $\{0,7\}$ is another. In fact, any advance that is relatively prime to the order of the group will do. Here $1,5,7,11$ are relatively prime to 12. Some restriction on the generators is needed to get the 12 elements of the clock group using addition. Without a restriction we would generate all the natural numbers by repeatedly adding 1. The restriction

$$1^{12} = 0,$$

[4]An equally valid group can be defined with identity 12.

will do the trick. It simply says that adding 1 *twelve times* returns us to the original member of the group. Notice that we have used notation that suggests multiplication to represent relations between the members of the group even though the relation in the example resembles addition. An element that produces the identity when repeated $n$ times is said to have a *period* of $n$. For the generator set $\{0,7\}$, the equivalent restriction is $7^{12}=0$. A set of generators and an appropriate set of restrictions is called a *presentation* or an *abstract definition* of a group. The elements $\{0,1\}$ and the relation $1^{12}=0$ are an abstract definition of $G_{12}$.

A *subgroup* is simply a subset of the elements of a group that satisfies the group definition. In clock arithmetic the even elements including the identity form a subgroup. The elements $\{0,2\}$ and the relation $2^6=0$ are an abstract definition of subgroup $G_6$. It can be seen as a one-sixth rotation and no rotation. The group captures the structure of a transit schedule in which buses arrive at one stop exactly on every even hour and arrive at the next stop exactly five minutes past every even hour. Sets of related games can often be described by subgroups, and the same subgroup can describe relations among several different sets of games.

The concept of the *direct product* of two groups is also helpful in describing the 2 × 2 games. The direct product of two groups, $H_x$ and $F_y$, which have no common members except the identity is written $G_{xy} \equiv H_x \times F_y$. $H_x$ and $F_y$ are subgroups of $G_{xy}$.

If we define

$$H_4 \equiv \{0,3|3^4=0\} = \{0,3,6,9\}$$
$$F_3 \equiv \{0,4|4^3=0\} = \{0,4,8\}$$

then

$$H_4 \times F_3 = \{0,1,2,3,4,5,6,7,9,10,11\} = G_{12}$$

In other words, the direct product of these subgroups of $G_{12}$ is the entire group. The element 2, which appears in neither $F_3$ nor $H_4$, is part of $H_4 \times F_3$ because $8 \in F_3$, $6 \in H_4$ and $6+8=2$.

For the 2 × 2 games, the elements of the groups are transformations and the binary operation is the concatenation of two transformations. This means applying one transformation to a game and then applying the second transformation to the gamc resulting from the first. The basic swap operations are the generators of the groups we examine.

## 3.4 Constructing the graph of $2 \times 2$ games

Three hierarchical concepts are useful: *groups* with their subgroups, *graphs* with their subgraphs, and topological *spaces* with their subspaces.

- The set of six swaps is associated with a mathematical group that can be identified with the $2 \times 2$ games[5].
- Games can be treated as the vertices of a graph. The six swaps are then identified with the six edges that meet at each vertex.
- The entire graph consisiting of the 144 games and the links between nearest neighbours is a representatation of a topological space.

Removing elements from a set of generators (i.e. removing a swap) removes edges from the graph[6]. With one important and recurring exception, restricting the set of operators partitions the $2 \times 2$ games into a set of identical subspaces with identical graphs[7]. Among the games in each of these subspaces, payoffs vary in a simple and restricted way.

Topological features can be deduced from either the group structure or from the graph structure. We generally work from the graph or subgraph, but at each stage we describe the associated groups. We also present some topological features, but save most of the details for later chapters.

[5]If we use all six swaps the group is actually the 576 element bipermutation group $S_4 \times S_4$. This is precisely the set of ordinal bi-matrices in Figure 2.5 on page 16. The 144 game graph represents a complex of the group, but is not a subgroup.

[6]Removing any one of $R_{12}, C_{12}, R_{34}, C_{34}$, does not partition the graph of games. This implies that these four swap operators are not independent.

[7]The subspaces have identical structures, but they contain different sets of games.

### 3.4.1 Subgraph/subspace/subgroup generated by a single swap: $Z_2$

We begin with the simplest case – a single swap operator. Applying the operator to any game generates a cycle of two games. We then add an operator to include more games and generate a more complex graph. We continue adding until we have a graph that includes all the ordinal 2 × 2 games. This stepwise procedure systematically unfolds the structure of the graph.

The operation $R_{12}(g_{111})$ exchanges the values in bold type in the matrix in Figure 3.1.

$$R_{12}(g_{g_{111}}) = g_{121}$$

Applying $R_{12}$ to $g_{121}$ does not produce a third game: instead it returns us to the original game.

$$R_{12}(R_{12}(g_{g_{111}})) = R_{12}(g_{g_{121}}) = g_{111}$$

Each of the six swap operators completely partitions the set of 144 games into 72 non-overlapping 2-game subsets that are closed under the single operation. These sets are topological spaces under the Hausdorff axioms, and are subspaces of the larger topological space of 2 × 2 games. The pairs of games may be represented by two points, or vertices, joined by two directed edges. When an operation is its own inverse it is conventional to replace the two directed edges with a single undirected edge. Each of the six swap operators combines with the identity operator to produce a specific 2-element group. Each of these 2-element cyclic groups is an instance of the general group $Z_2$.

Table 3.2 summarizes what we know about the two-game subspaces produced using single swaps.

| *number of swaps* | *swap type* | *number of versions* | *number of games* | *number of sub-spaces* | *description* | *group* |
|---|---|---|---|---|---|---|
| 1 | any | 6 | 2 | 72 | edge | $Z_2$ |

Table 3.2: Two-game cycles

### 3.4.2 Two non-overlapping operations: $Z_2 \times Z_2$

We call swap operations that do not affect the same rank payoff for a given player *non-overlapping* operations. The operations $C_{12}$ and $C_{23}$ overlap since both affect the second-ranked outcome for the column player. $C_{12}$ and $C_{34}$ do not overlap. $C_{12}$ and $R_{12}$ cannot overlap because they operate on the payoffs for different players. We call operations on the payoffs of different players *orthogonal*. Concatenating equivalent orthogonal swaps yields a new game:

$$R_{12}(C_{12}(g_{111})) = R_{12}(g_{112}) = g_{122}$$

The new game, two swaps from the Prisoner's Dilemma, is Chicken. Repeating the combination returns the Prisoner's Dilemma:

$$\begin{aligned} R_{12}(C_{12}(R_{12}(C_{12}(g_{111})))) &= R_{12}(C_{12}(g_{122})) \\ &= R_{12}(g_{121}) \\ &= g_{111} \end{aligned}$$

Concatenating any pair of orthogonal swaps connects the games into sets of four. These 4-game sets are also subspaces of the entire space of the $2 \times 2$ games.

There are nine orthogonal pairs of swaps. Two additional non-orthogonal but non-overlapping pairs result from combining $C_{12}$ and $C_{34}$ or $R_{12}$ and $R_{34}$. There are therefore 11 distinct ways to partition the $2 \times 2$ games into four-game cycles. These cycles are the main building blocks of the larger structure of the $2 \times 2$ games. They form *faces* of the polytope that appears when the full topology is laid out.

> **Face:** The intersection of an n-dimensional polytope with a tangent hyperplane. Zero-dimensional faces are known as polyhedron vertices (nodes), one-dimensional faces as polyhedron edges.

Each non-overlapping pair of swaps combined with the identity element, for example $\{C_{12}, R_{12}, I\}$, generates a 4-element group. The group is the direct product of two 2-element groups. The restrictions are expressed in these relations.

$$C_{12}^2 = R_{12}^2 = [C_{12}R_{12}]^2 = I$$

One set of orthogonal swaps is of particular interest. We call the subspaces produced by combining $C_{12}$ and $R_{12}$ *tiles*. The 36 tiles consist of groups of games that are closely related in an economic sense. $C_{12}$ and $R_{12}$ permute the lowest- and second-lowest ranked outcomes, which are the swaps least likely to change the decisions of players concerned to maximize their payoffs. Nash equilibria require that players choose the highest payoff within their inducement correspondences. In 21 of 36 tiles the equilibrium payoffs are the same for every game. In 12 tiles the payoff for one player changes and in only three are payoffs changed for both players. Of these three, there is only one maximally diverse tile, containing four games with different equilibria. That unusual tile is the one containing the Prisoner's Dilemma.

| *number of swaps* | *swap type* | *number of versions* | *number of games* | *number of sub-spaces* | *description* | *group* |
|---|---|---|---|---|---|---|
| 2 | non-overlapping | 11 | 4 | 36 | face (tile) | $Z_2 \times Z_2$ |

Table 3.3: Four-game cycles

### 3.4.3 Overlapping operations: $P_6$

There are four groups generated by *overlapping* pairs of swaps: $\{C_{12}, C_{23}, I\}$, $\{C_{23}, C_{34}, I\}$, $\{R_{12}, R_{23}, I\}$, and $\{R_{23}, R_{34}, I\}$. Like non-overlapping pairs, the overlapping pairs produce closed subsets of games, but they produce a different pattern from the non-overlapping pairs. To see why, look at the effect of alternately swapping first the 1 and the 2, then the 2 and the 3 beginning with the sequence 1234:

$$1234 \leftrightarrow \mathbf{21}34 \leftrightarrow \mathbf{3}1\mathbf{2}4 \leftrightarrow 3\mathbf{21}4 \leftrightarrow \mathbf{23}14 \leftrightarrow \mathbf{1}3\mathbf{2}4 \leftrightarrow 1\mathbf{23}4$$

The effect is to run through the permutations of the numbers $1, 2, 3$ without changing the position of 4.

Overlapping operations are not commutative. For example,

$$C_{23}(C_{12}(g_{111})) = C_{23}(g_{112}) = g_{113}$$

but

$$C_{12}(C_{23}(g_{111})) = C_{12}(g_{116}) = g_{115}.$$

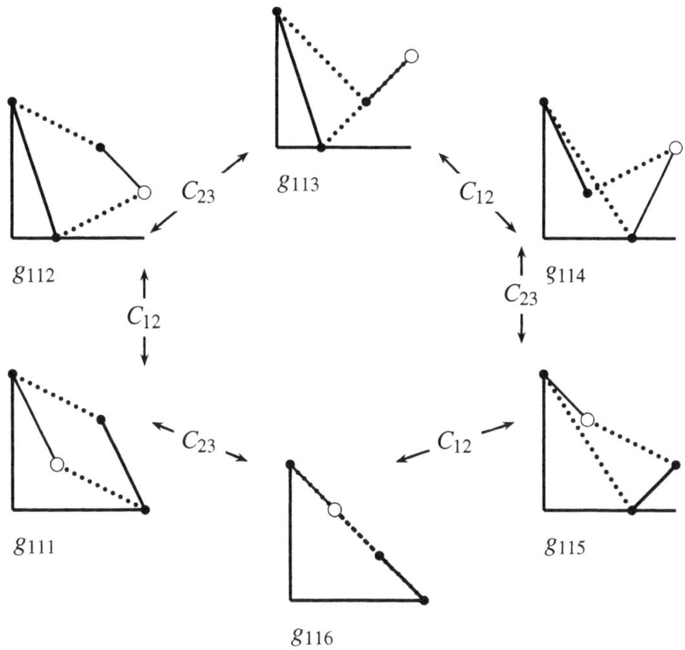


Figure 3.3: Games generated by $C_{12}$ and $C_{23}$ from $g_{111}$

Starting with $g_{111}$, repeated application of $C_{12}$ followed by $C_{23}$ generates the 6-game cycle $g_{111}$, $g_{112}$, $g_{113}$, $g_{114}$, $g_{115}$, $g_{116}$. See Figure 3.3.

To simplify notation, let

$$C_{23}(C_{12}(g)) \equiv C_{23}C_{12}(g)$$

and let

$$C_{23}(C_{12}(C_{23}(C_{12}(g)))) \equiv [C_{23}C_{12}]^2(g).$$

It is easily verified that

$$[C_{23}C_{12}]^3(g) \equiv C_{23}(C_{12}(C_{23}(C_{12}(C_{23}(C_{12}(g)))))) = \mathrm{g}$$

which is to say, repeating the swaps $\mathrm{C}_{12}$ and $\mathrm{C}_{23}$ three times results in a closed loop of six games. The group generated by $C_{12}$ and $C_{23}$ (or analogous overlapping pairs) has six elements and the combined operation $C_{12}C_{23}$ is of period three. The abstract definition using the generators $\{C_{12}, C_{23}, I\}$ is

$$C_{12}^2 = C_{23}^2 = [C_{12}C_{23}]^3 = I \tag{3.1}$$

which is the *symmetric permutation group* of six elements, $P_6$. This group divides the $2 \times 2$ games into 24 cycles of six games. The games in this cycle share two properties. First, the payoffs for the row player are unchanged since only column swaps have been used. Second, the location of the highest payoff for the column player is fixed.

We can construct similar cycles using $C_{23}$ and $C_{34}$ where the 3 payoff is manipulated by both swaps. The $X_{34}$ swaps differ from $X_{12}$ swaps in that they permute the preference ordering over the three most preferred outcomes. The resulting cycles consist of games in which the location of the lowest payoff for the column player is invariant. The $\{C_{12}, C_{23}, I\}$ cycles contain games more closely related than games in $\{C_{23}, C_{34}, I\}$ cycles: the payoffs most likely to form a Nash equilibrium are least likely to be affected in the former and most likely to be affected in the latter. We have used this observation in deciding the indexing system for the $2 \times 2$ games. Column indices enumerate the elements of subsets based on $C_{12}$ and $C_{23}$. In figure 2.8 on page 20, these are the 24 rows. Row indices enumerate the $\{R_{12}, R_{23}, I\}$ subsets and identify columns in the figure.

| *number of swaps* | *swap type* | *number of versions* | *number of games* | *number of sub-spaces* | *description* | *group* |
|---|---|---|---|---|---|---|
| 2 | over-lapping | 4 | 6 | 24 | loop (column or row) | $P_6$ |

Table 3.4: Six-game cycles

### 3.4.4 Slices: $P_{24}$

If we now add $C_{34}$ to the generators $\{C_{12}, C_{23}, I\}$ that produce the loop row, we create a symmetric permutation group, $S_4$ of $4 \times 3 \times 2 = 24$ elements. There is an equivalent group using row operations.

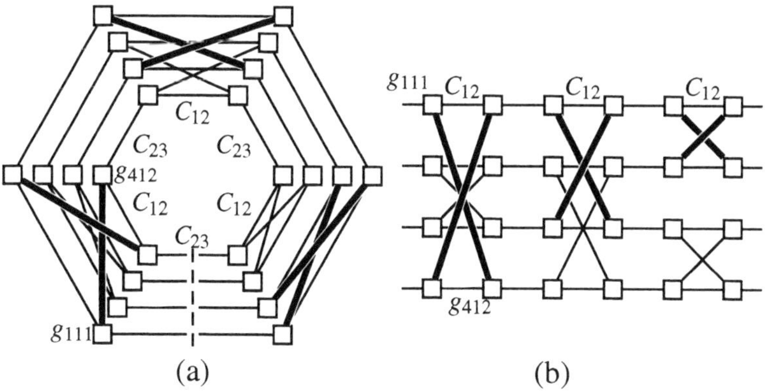


Figure 3.4: Slice

Each permutation group distinguishes six subspaces which we call *slices*. A slice contains the games with a fixed payoff pattern for one player paired with all possible payoff patterns for the opponent.

In Figure 3.4(a), the graph for a $\{C_{12}, C_{23}, C_{34}, I\}$ slice appears as four interlocked loops. Each loop represents a separate row of games. The pattern is the same for all slices produced using the set of column swaps or the corresponding row swaps.

In Figure 2.8, this slice consists of the first rows of the four layers. Figure 3.4(b) shows the connections between the rows by cutting the loops of 3.4(a) through four radially aligned $C_{23}$ links. We move from layer to layer with $C_{34}$ swaps, for example from $g_{111}$ in the top layer to the second game in the bottom layer, $g_{412}$. Notice that if two layers are linked by the $X_{34}$ swap, then the other two layers are also linked to each other.

Slices have a useful interpretation in terms of restrictions on information. If the row player knows her own payoff structure, but nothing about the payoff pattern for the column player, there are exactly 24 strict ordinal combinations that she might be facing. Combined with her own known payoff, each represents a possible game. The games in a slice are the games that she might be in, given only information about her own payoffs.

| *number of swaps* | *swap type* | *number of versions* | *number of games* | *number of sub-spaces* | *description* | *group* |
|---|---|---|---|---|---|---|
| 3 | C or R | 2 | 24 | 6 | slice | $P_{24}$ |

Table 3.5: Twenty-four game slices

### 3.4.5 Structure of a stack

The $C_{34}$ swap always leads to a game on another layer. $C_{34}$ combined with $C_{12}$ (or $R_{34}$ with $R_{12}$), leads to a game directly "over' or "under" the initial game. The new game shares the same row and column indices. In the representation we propose, the game reached by this combined operation is said to be in the same *stack*.

The payoff matrices in a stack are all constructed from the same payoff pattern for each player. On each layer, the relative orientation of the payoffs is different. The $g_{\ell 11}$ stack containing Prisoner's Dilemma is shown in Figure 3.5 in the next section.

### 3.4.6 Layers: $P_6 \times P_6$

In Chapter 2 we introduced the *layer* as a convenient organizing structure for the 2 × 2 games. A layer is the direct product of the row group and the column group. These orthogonal loops are cyclic groups of order 6 which have only the identity element in common. The direct product therefore exists and is of order 36. The group corresponds to a class of subspaces of 36 games.

The graphs of layers are regular 4-connected point lattices, since every game has four neighbours in the subspace. To establish that the lattice is a simple surface, notice that every game is one of six forming a closed loop with $\{C_{12}, C_{23}, I\}$. The same argument using row swaps shows every game will also be a member of another closed loop of six games. Any loop produced by column swaps therefore intersects six transverse loops generated by row swaps. To see that the transverse loops are joined into a surface, recall that non-overlapping pairs commute: $C_{ij}R_{kl}$ yields the same game as $R_{kl}C_{ij}$. This says that, for any two nearest neighbours in a row, the nearest neighbours under $R_{kl}$ are themselves nearest neighbours in a row. It follows immediately that the transverse loops form a grid consisting of 36 games.

A layer of 36 games is most conveniently represented as a $6 \times 6$ grid as in Figures 2.13 to 2.16 on pages 30 and 31. For each game in these subspaces, four of six possible neighbours are in the subspace[8]. Moving from left to right across each layer, the operators $C_{12}$ and $C_{23}$ alternate, beginning in our preferred arrangement with $C_{12}$. $R_{12}$ and $R_{23}$ alternate from bottom to top.

| *number of swaps* | *swap type* | *number of versions* | *number of games* | *number of sub-spaces* | *description* | *group* |
|---|---|---|---|---|---|---|
| 4 | over-lapping pairs | 4 | 36 | 4 | layer | $P_6 \times P_6$ |

Table 3.6: Thirty-six game layers

## The "no-conflict" layer

The games in a layer have a common feature: the operations that generate a layer, $\{C_{12}, C_{23}, R_{12}, R_{23}, I\}$, leave the position of the payoff 4 unchanged for both players. Figure 3.5 shows the four patterns that are possible. The payoff matrix may have the two 4s in the same cell (Layer 3), in cells that are diagonally opposite (Layer 1), in two cells in the same row (Layer 2), or in two cells in the same column (Layer 4).

The location of the highest payoffs gives each layer a special character. Layer 1 is confrontational while games on Layer 3 tend to consensus. On Layer 2 the best outcomes for both players are at the two ends on an inducement correspondence for the column player. Since Column gets to choose, Column tends to do well on Layer 2. On Layer 4 Row has the advantage.

One of these patterns has been recognized by previous writers. Rapoport and Guyer [23], Rapoport, Guyer and Gordon [24], and Brams [5] describe the 36 games with the payoff combination (4,4)

[8] The other two neighbours for each game result from swap operations $X_{34}$ that do not appear on these surfaces.

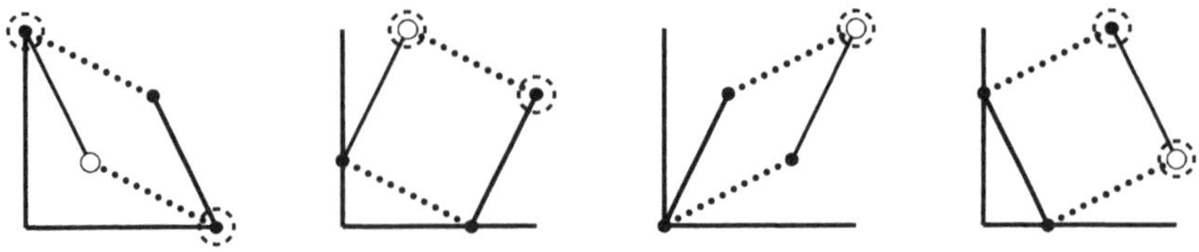

Layer 1: $g_{111}$ Layer 2: $g_{211}$ Layer 3: $g_{311}$ Layer 4: $g_{411}$

Figure 3.5: The stack and the position of highest payoffs

on layer 3 as "*no-conflict games*". In the Rapoport, Guyer and Gordon typology these games constitute a "phylum", which is the highest level in their classification. The phylum clearly has some topological basis.

## 3.4.7 Topology of a layer

The cyclic property of the row and column groups produces the topology of the layers. Games at the left edge of each layer in Figures 2.13 to 2.16 are neighbours of the games in the same row on the right edge. Games at the top edge of each layer are neighbours of the games in the same column on the bottom edge. To show this, in Figure 3.6 we roll one of the layers to form a cylinder. Since the games at the top and bottom are also neighbours, the cylinder must then be stretched and bent so that its ends meet. This procedure for mapping the games onto a torus has rows passing through the hole and columns encircling it[9]. Each layer forms a torus.

## 3.4.8 The Euler – Poincaré characteristic

One of the earliest results in mathematical topology, the *Euler – Poincaré characteristic*, or Euler number, is a computation that determines whether a particular graph can be drawn on a plane surface or sphere without crossing lines. If a graph has an Euler number of *two*, it can be drawn without crossing edges on a sphere. If the Euler number is *zero*, it can be drawn without crossing edges on a torus but

[9]Orthogonal cyclic operations of order three or more always generate a torus.

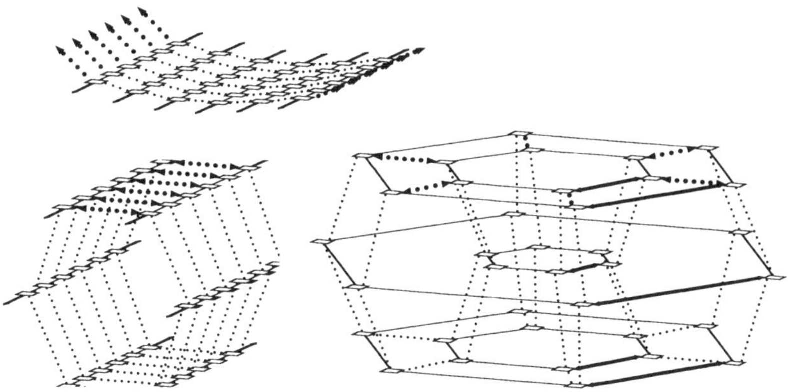

Figure 3.6: Torus

not on a sphere. A doughnut with two holes has an Euler number of *negative two*. Each additional hole adds −2 to the Euler number.

For a given graph, the Euler number is computed by adding the number of vertices to the number of faces and subtracting the number of edges.

$$Euler\ number = V + F - E$$

For the 36-game surface of one layer, each game is a vertex and the transformations are edges. Faces have four edges. Vertices, edges and faces can be counted in Figure 3.6. The surface contains 36 vertices. It has $4 \times 36 \div 2 = 72$ edges, since every edge is shared with one other game. Every game is adjacent to 4 faces and every face has four vertices, so the number of faces is 36. The Euler number is then $V + F - E = 36 + 36 - 72 = 0$, confirming that the 36-game subspaces are toruses.

### 3.4.9 The four-layered torus

The four layers of 36 games can be seen as four nested toruses. The four layers are linked by the $C_{34}$ and $R_{34}$ swaps into a more complex surface. The pattern of links was introduced in Section 3.4.4 where we defined the slice. The complete topology is explored later but in the next sections we briefly describe the "tiling" of the layers and the kind of structures that join the layers together.

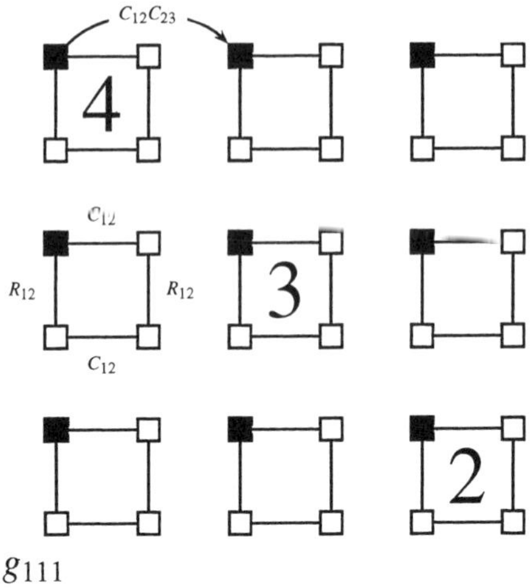


Figure 3.7: Tiles on and links from Layer 1

### 3.4.10 Tiling the layers

The 36-game layers generated by $\{C_{12}, C_{23}, R_{12}, R_{23}, I\}$ partition into nine *tiles* of four games each based on $\{C_{12}, R_{12}, I\}$. The $X_{23}$ swaps are removed from the set of generators so the surface breaks up into the nine pieces shown in Figure 3.7. Gaps between tiles stand for $X_{23}$ swaps.

If we define a new set of generators that includes combined "12" and "23" swaps,

$$\{R_{12}R_{23}, C_{12}C_{23}, I\},$$

with the restrictions $(R_{12}R_{23})^3 = (C_{12}C_{23})^3 = I$, we can generate 9-game subspaces consisting of, for example, the upper left game on each tile. The dark squares in Figure 3.7 show the location of this subspace. The direct product of this subgroup of order 9 and the tile group of order 4 generated by $\{I, C_{12}, R_{12}\}$ is once again the layer of order 36.

### 3.4.11 Pipes and hotspots

The four-layered torus provides a model for visualizing how the games are related. Swaps $C_{34}$ and $R_{34}$ "stitch" the layers together.

For each $X_{12}$ swap that joins a game to a neigbour on the same layer, there is a corresponding $X_{34}$ swap that connects it to a neighbour on another layer.

The numbers 2, 3, 4 in Figure 3.7 are the key to locating the links between layers. Each number shows which layer is connected to Layer 1 along a row and a column of tiles.

For example, say we want to find the neighbours of $g_{111}$ on other layers. We need to know which layer is linked to Layer 1 between Rows 1 and 2. The "2" in on the lower right tile tells us that games in Row 1 on Layer 1 are row-connected to games on Layer 2. (It follows that games on Layer 2 are row-connected to games on Layer 1 and that games on Layer 3 are connected to Layer 4.) The row neighbour of $g_{111}$ is therefore $g_{221}$.

The "4" on the upper left tile tells us that between *Columns* 1 and 2, games on Layer 1 are connected to games on Layer 4. The column neighbour on another layer is therefore $g_{412}$[10].

If $R_{34}$ and $C_{34}$ swaps are added to the set of tile generators, $\{R_{12}, C_{12}, I\}$, the resulting set $\{R_{12}, C_{12}, R_{34}, C_{34}, I\}$ generates two distinct groups with different restricting relations. These swaps do not partition the $2 \times 2$ games into equivalent subspaces. Instead, we get new important classes of objects called *pipes* and *hotspots* which are piles of tiles linked by $X_{34}$ swaps. We will discuss pipes and hotspots in detail in Chapters 6 and 7.

Hotspots are composed of two tiles and occur when $R_{34}$ and $C_{34}$ link to the same layer. Pipes which are composed of four tiles occur when $R_{34}$ and $C_{34}$ link to different layers. In Figure 3.7 the hotspots occupy the negative diagonal of the layers (the large digits). The four-tile, 16-game pipes are located at the other six tiles.

Each hotspot and pipe forms a subspace and is associated with a group. The fact that there are groups of different orders generated by the same set of swap operators is the most distinctive feature of the topology of the $2 \times 2$ games. This distinction remains in all subsets of the swaps that generate pipes and hotspots. It prevents the complete set of six swap operators from forming a group.

---

[10]The reader may want to return to Figure 3.2 and locate $g_{412}$ and its six neighbours using the information in Figure 3.7

| *number of swaps* | *swap type* | *number of versions* | *number of games* | *number of sub-spaces* | *description* | *group* |
|---|---|---|---|---|---|---|
| 4 | non-overlapping | 1 | 8 OR 16 | 12 | Hotspot OR Pipe | $Z_3$ OR $Z_4$ |

Table 3.7: Pipes and hotspots

## 3.5 Structure and content

To this point we have concentrated on developing a description of the topological space of the 2 × 2 games induced by the structure of preferences. The space can be partitioned into subspaces, and the subspaces can be represented as surfaces. We have shown for example that the layers are toroidal.

We have also shown that related games may be associated with distinct subspaces. An example is the class of "no conflict" games which occupies layer three of the topological structure. The next five chapters explore the topological relationships within the important subspaces of games. In each case, we show how the structure of subspaces is related to fundamental concepts of game theory such as Nash equilibria, dominance solvability and conflict.

# Chapter 4

# Symmetric games

## 4.1 The seven most studied $2 \times 2$ games

The 12 symmetric games provide models for social situations ranging from resource wars and marriage through hunting parties and office games. It is possible to discuss an astonishing range of issues in philosophy, biology and economics using only symmetric games.

A symmetric game is often the convenient representative of a collection of related games, most of which are asymmetric. For example, there is a nine-game region on Layer 1 consisting of games that resemble the Battle of the Sexes. Six are asymmetric. All nine have two efficient equilibria that favour different players, raising the possibility of distributional conflict. One of the the symmetric games in this region is Chicken, the game made famous by Bertrand Russell as a model for the nuclear arms race. Chicken is perhaps the second most familiar $2 \times 2$ game.

The Prisoner's Dilemma, which is adjacent under a symmetric transformation to Chicken on Layer 1, is certainly the most famous of all 144 $2 \times 2$ games. It is one of only seven with a unique and inferior Nash equilibrium and the only symmetric member of the family. We devote all of Chapter 5 to the Prisoner's Dilemma and its relatives.

There is also a nine-game region on Layer 3 that includes two symmetric versions of the Coordination game. Each game in the region has two equilibria, one of which is Pareto inefficient. The symmetric game Stag and Hare, also known as Stag Hunt, and "the meeting game", lies in this region.

In total, seven of the most studied $2 \times 2$ games are symmetric: the Prisoner's Dilemma, Chicken, Stag and Hare, the two Coordination games, and the two versions of the Battle of the Sexes (BoS).

This chapter explores relationships among the symmetric games, partly because they are an important topic in their own right, and partly because they put the topological approach to work on a relatively simple subspace of the $2 \times 2$ games. The exercise yields several basic results about the topology of the $2 \times 2$ games.

## 4.2 The nature of a symmetric game

The symmetric games are used so often, especially in introductions to game theory, that it is easy to forget they represent a very special case. For each strategy of the row player in a symmetric game, there must be an equivalent strategy for the column player. The strategies' names may not make the equivalence obvious. We use $U$ and $D$ for Row's strategies and $L$ and $R$ for Column's to take advantage of the familiar directions in the bi-matrix structure. Many discussions of symmetric games name the alternatives "Defect" and "Cooperate" for both players, to emphasize that they are in a symmetric situation.

If both players have strategic choices $x$ and $y$, payoffs for the symmetric games are restricted in a simple but very strong way:

$$\mathrm{r(x, y) = c(y, x)} \tag{4.1}$$

In this equation, $r(x,y)$ is the payoff to Row if she chooses strategy $x$ when Column chooses $y$, and $c(y,x)$ is the payoff to Column if Row chooses strategy $y$ when Column chooses $x$.

Condition 4.1 restricts the payoff matrix to the form shown in the upper left in Figure 4.1. Figure 4.1 also shows the matrix order graph for game $g_{144}$, one of the symmetric Battles of the Sexes[1]. Notice that the highest symmetric outcome is in the upper right of the payoff matrix, following the convention introduced on page 17.

[1]Like every symmetric game, the index for $g_{144}$ begins with a 1 or a 3, followed by a pair with the same value. The last two numbers in the index are the same because the symmetric games lie on the positive diagonals of layers one and three.

| $L$ | $R$ |
|---|---|
| $D,C$ | $\mathbf{B,B}$ |
| $\mathbf{A,A}$ | $C,D$ |

| $L$ | $R$ |
|---|---|
| 4,3 | 2,2 |
| 1,1 | 3,4 |

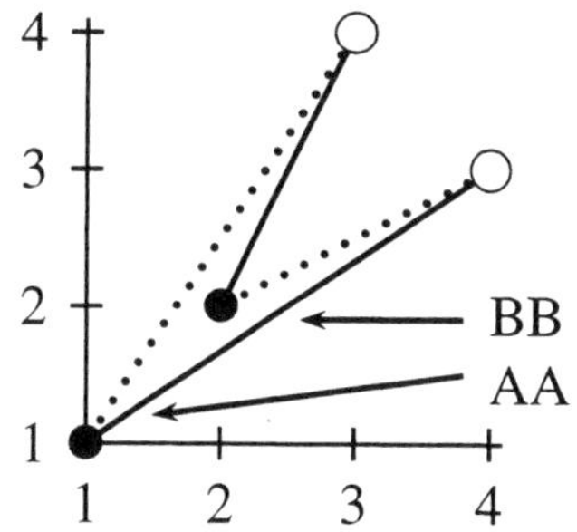


Battle of the Sexes, $g_{144}$

Figure 4.1: Symmetric games

## 4.3 Counting the symmetric games

Counting the symmetric games is our first task. This section presents a method that leads directly to the topological relationships among the games. We then use the topological relationships to describe features of specific symmetric games. The approach combines aspects implied by the definition with the method presented in Section 2.4.1 to count the 144 $2 \times 2$ games.

Condition 4.1 allows for two cases, $y = x$ and $y \neq x$ . The first describes the two ways the players can choose the same strategy. In those cases they get the same payoff, so there are two payoff pairs with the same rank for both players. Since we are dealing with strict ordinal games, we can arbitrarily pick ranks $A$ and $B$ satisfying $A < B$. The second case describes two ways the players can choose different strategies. There is really only one combination left to choose since, if the third point is $(C,D)$, the fourth must be $(D,C)$.

Counting the ways that $(C,D)$ can be chosen is equivalent to counting the symmetric games. Since $C \neq D$, and $C,D \in \{1,2,3,4\}$, the number is simply ${}_4P_2 = 4 \times 3 = 12$. The other two values are then $A$ and $B$.

A geometric approach is more revealing. The value of $C$ must fall below $B$, between $B$ and $A$, or above $A$. Since a point can fall in any of three horizontal intervals and any of three vertical intervals, it appears there are nine regions to consider with two dimensions, as

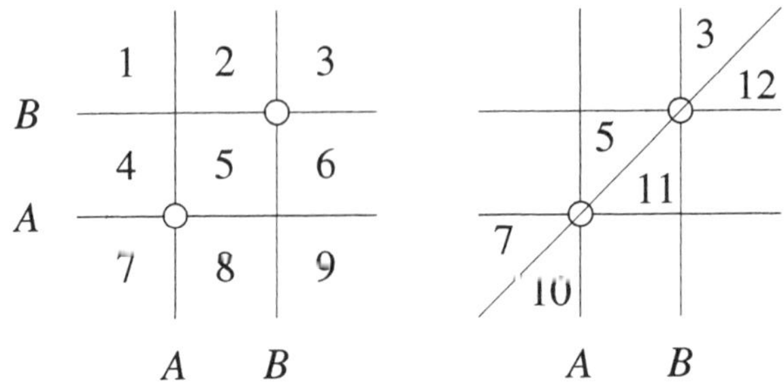


Figure 4.2: Counting the symmetric games geometrically: steps 1 and 2

illustrated on the left of Figure 4.2. There is a small complication, however. Even if we know, for example, that both the row and column values are less than A, we still don't know if the row value is greater or less than the column value. Fortunately we know that the row value is higher than the column value everywhere to the right of a positive diagonal through the origin. To make this distinction we add a diagonal line to the figure on the right.

In the right panel of Figure 4.2, "10", "11" and "12" mark points where the row value is higher than the column value. The "3", "5" and "7" illustrate the opposite case. Introducing a diagonal separates these cases, yielding twelve regions. Each corresponds to a game.

### 4.3.1 Identifying the symmetric games

Every ray originating from the upper symmetric payoff and ending in one of the 12 regions identifies a symmetric game. Each region represents one way of choosing two numbers from four. The two numbers are the ranks for the two players for one of the non-symmetric outcomes. Adding the rank information to Figure 4.2 produces Figure 4.3. These ordered pairs can serve as *identifiers* for the symmetric games. A symmetric game is completely determined by either

1. connecting the upper symmetric point to a point in any one of the 12 regions with a solid line, or

2. placing the corresponding pair of rankings from Figure 4.3 in the lower right cell of a payoff matrix.

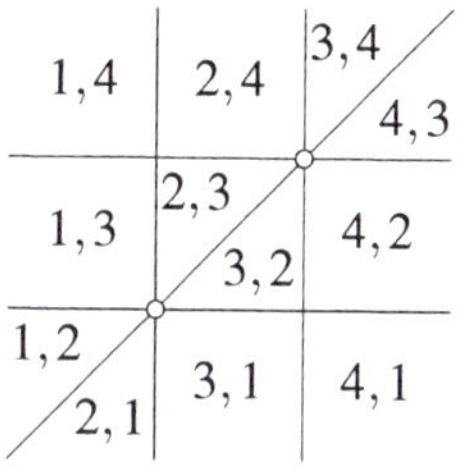


Figure 4.3: Identifiers for the symmetric games

Positioning a solid line in the order graph and specifying the lower right cell of the payoff matrix are ways of saying that we are fixing the inducement correspondence for the row player.

### Example

The payoff pair $(4,1)$ is the identifier for the Prisoner's Dilemma. To construct the order graph, remember that $(4,1)$ is in Row's right inducement correspondence, i.e. the one that includes the upper symmetric point. Since 4 and 1 are used in the identifier, the two symmetric pairs must be $(2,2)$ and $(3,3)$. The solid line is drawn for $(3,3)$ to $(4,1)$. The remaining point must be $(1,4)$. Constructing the Prisoner's Dilemma is illustrated as Figure 4.4. The right panel is a complete strategic form representation.

The payoff matrix for the Prisoner's Dilemma can be constructed by placing $(4,1)$ in the lower right of a $2 \times 2$ matrix with $(3,3)$ above, $(2,2)$ to the left and $(1,4)$ diagonally opposite.

Any symmetric game can be transformed into one of its neighbours in another ordinal class by dragging the identifier point across a boundary. The identifier is the tip of one "wing" of the quadrilateral in the order graph. Treating the wingtips symmetrically produces a symmetric neighbour. For example, dragging the lower wingtip of the Prisoner's Dilemma up into the right-centre cell, and the upper one into the top-centre cell produces Chicken, $g_{122}$[2]. The combination of swaps required to turn the Prisoner's Dilemma into

[2]When the Prisoner's Dilemma flaps its wings up farther we get the Battle of the Sexes and when it flaps its wings down, we get Stag Hunt and the Coordination games.

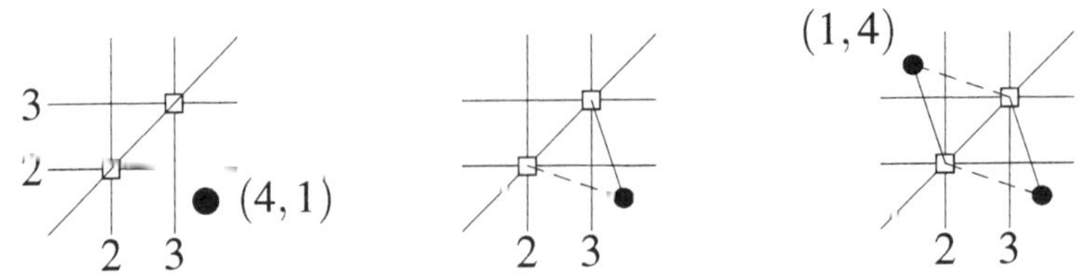


Figure 4.4: Constructing the PD from two numbers

Chicken is $C_{12}$ and $R_{12}$, which we write $S_{12}$. The identifier for Chicken is (4, 2). See Figure 4.5.

## 4.4 The space of symmetric games

The symmetric games form a proper subspace under the three symmetric operations $S_{12}$, $S_{23}$, and $S_{34}$. The links created by these operations are shown in Figure 4.6. The figure presents a great deal of information about the symmetric games and illustrates structural features of the entire collection of games.

The 12 symmetric games consist of four isometries of just three basic payoff configurations. There are only three because symmetric games must have exactly two payoff points on the positive diagonal, eliminating four of seven possible elementary payoff patterns. The four points must also be connected symmetrically, further eliminat-

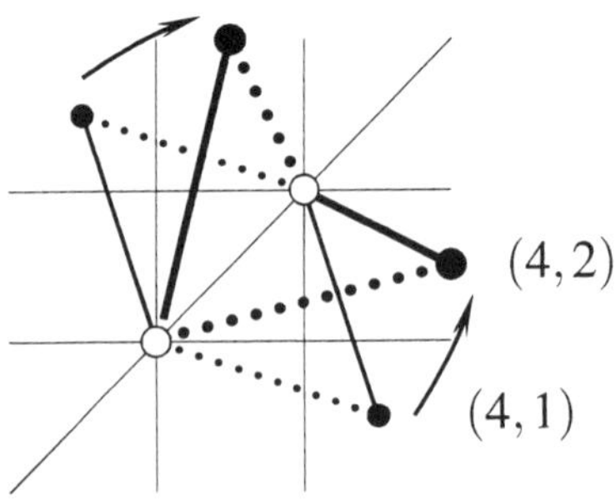


Figure 4.5: Turning the PD into Chicken

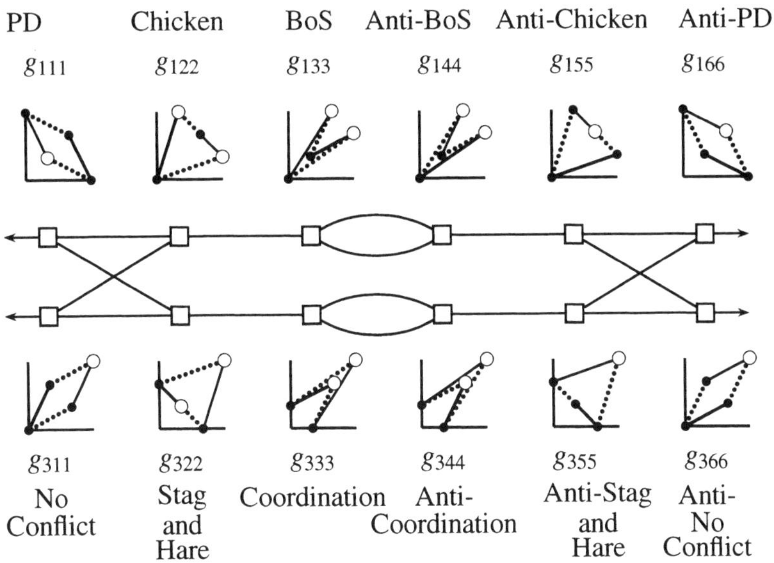


Figure 4.6: The linked-loop model of the symmetric games

ing half of the games built on each of the three patterns remaining. There are only four symmetric isometries of each pattern because the presence of an axis of symmetry makes some reflections equivalent.

Two pairs of games are doubly linked in Figure 4.6. The symmetric swap $S_{34}$ usually results in a move between layers 1 and 3. Between $g_{133}$ and $g_{144}$, and between $g_{333}$ and $g_{344}$, something different happens. $S_{34}$ transforms these four games to the same neighbour as $S_{12}$ does. This is a pattern that will be repeated with other subsets of games.

The doubly linked games in the top layer are the two symmetric versions of the BoS. The doubly linked games in layer three are symmetric versions of the Coordination game[3]. These games have only

[3]The doubly linked games are clearly distinct, but we know of no one who has drawn attention to the differences between the two symmetric BoS games or between the two versions of the Coordination game. Without strict ordinality, the BoS may appear with the two symmetric points in the same place, possibly (0,0), and the Coordination games may have their two asymmetric points in the same place.

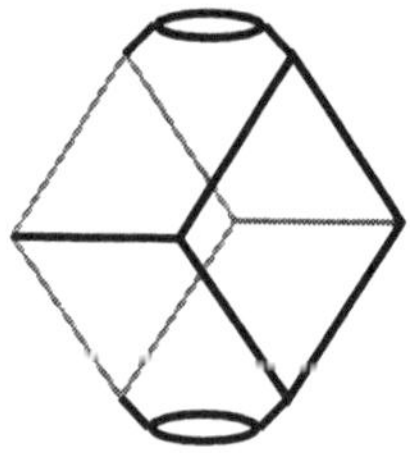

Figure 4.7: Symmetric games as vertices of a polyhedron

two distinct neighbours. In an interesting sense, the doubly linked games are closer to each other than most adjacent games. Each fills two thirds of the other's symmetric neigbourhood.

The doubly linked *pairs*, on the other hand, are farther from each other than any other symmetric games – it takes four or five symmetric operations to get from a Coordination game to a Battle of the Sexes. All other pairs of symmetric games are no more than three symmetric swaps apart. The double linked pairs can be seen as "poles" of the subspace of symmetric games.

The linked-loop model in Figure 4.6 contains all the information we need about the topology of the symmetric subspace, but there are other revealing representations. In Figure 4.7 we continue to treat the games as the nodes of the graph, as in Figure 4.6, but we close the loops and untwist the crossed lines. The graph is like a cube with circles inserted into a pair of opposite edges. In this form it is clear that the graph of the symmetric games can be mapped onto a sphere. The circles can be seen as the Arctic and Antarctic circles, at once polar opposites and reflections of each other.

The double linkage results from the one non-uniformity in the topology of the $2 \times 2$ games. It occurs whenever $R_{34}$ and $C_{34}$ lead to the same layer. When there are two orthogonal operations that must each link to one of the three other layers there are nine possible combinations. Inevitably, three pairings must reach the same layer.

In our standard configuration, double links occur only on the main negative diagonal of the four layers. We examine the feature that causes double links in Chapter 6. Here we are only interested in the effect it has in the subspaces generated by symmetric operations.

## 4.5 A map of the symmetric games

In Figures 4.6 and 4.7, the symmetric games are nodes of a graph. Treating games as nodes fits our approach which uses ordinal games as representatives of regions in the space of real-valued $2 \times 2$ games. An alternate representation, the *dual* in which game vertices appear as faces of a polyhedron, is particularly useful for exploring real-valued variants of a given game, as we do in Chapter 10. Each face represents an equivalence class in the continuous space of $2 \times 2$ games associated with one ordinal game.

To produce the polyhedral version in Figure 4.8 the appropriate order graphs are inserted in each region of Figure 4.2. The result is a two-dimensional map of the space of the symmetric games.

Figure 4.8 is perhaps the most useful representation of the symmetric $2 \times 2$ games. It can be constructed easily beginning from two pairs of crossing lines and a diagonal. The game matrix and the order graph can be read directly from the figure or it can be used to construct symmetric games with specific properties. The figure shows which games are neighbours, and it divides naturally into meaningful sub-regions.

### 4.5.1 Types of symmetric games

The most interesting sub-regions are shown in a sequence of thumbnails at the bottom of the figure. Games above the positive diagonal in Figure 4.8, for example, are all "anti-" games. Each is an $R^{\nwarrow}$ reflection of a game below the diagonal.

The six games in the lower left, including two pairs on the diagonal, are from Layer 3 which includes all games with the possibility of a $(4,4)$ payoff. Three of the games have two equilibria. They are all variants of the Coordination game.

The remaining six games are from Layer 1. Three of the games have two equilibria. They are all variants of the Battle of the Sexes, although $g_{122}$, known as Chicken, is generally seen as a distinct type.

The five games in the upper left corner of Figure 4.8 are simply boring. They have Nash equilibria that are unique, efficient, and unsurprising. They have attracted virtually no attention from analysts.

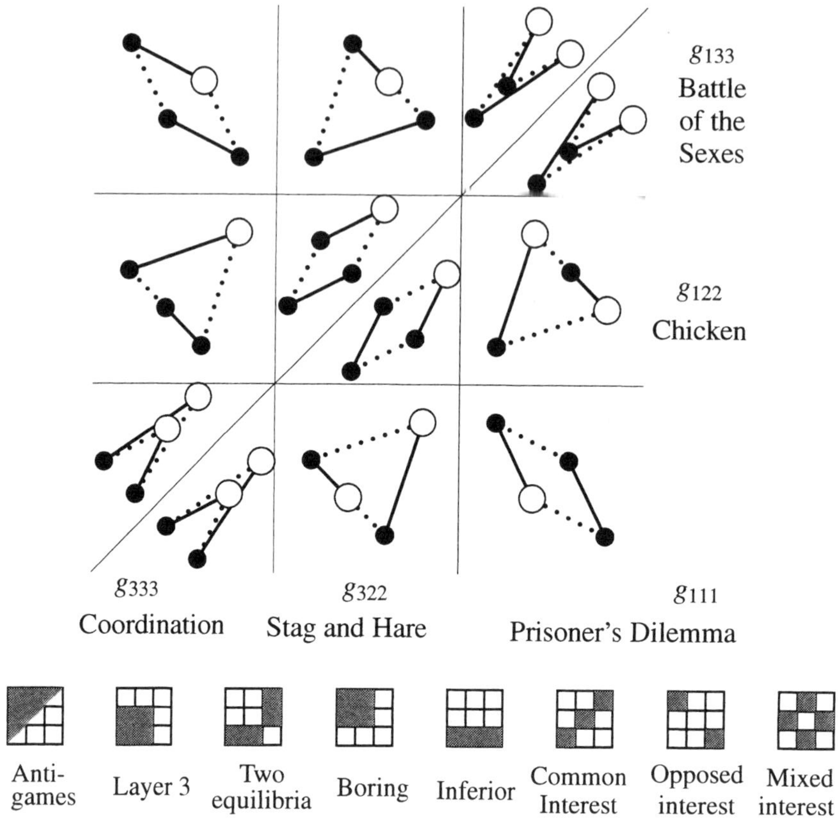


Figure 4.8: The space of the 12 symmetric games

Results presented in Chapter 10 suggest that at least some games in this region warrant further study.

The seven games in white that border the boring games are all interesting. Except for the Prisoner's Dilemma, all have two Nash equilibria. The Prisoner's Dilemma at the intersection has a unique Pareto-inefficient equilibrium.

There are six games in which the interests of the players are always aligned in that, for any inducement correspondence, both players prefer the same strategy choice. All inducement correspondences are positively sloped for the games adjacent to the positive diagonal of the figure.

There are only two games of strictly opposed interests: in every inducement correspondence, any strategy choice that benefits one player makes the other worse off. In these games all the inducement correspondences are negatively sloped. Of these, the PD has an Pareto-inefficient Nash equilibriun while the Nash equilibrium for the anti-PD is efficient.

The remaining four games, all isometries of Chicken, are games in which the interests of players are sometimes aligned and sometimes opposed. Games of conflict and common interest are investigated more fully in Chapter 8.

### 4.5.2 The world of the symmetric games: a flying octahedron

The two-dimensional map in Figures 4.8 can be deformed to cover a sphere. It is more revealing to construct a closed polyhedron in which each facet is an equilateral triangle. The resulting object looks like a winged octahedron, or a rugby football. The first step, at the top of Figure 4.9 is to complete the triangles around the games in the four corners. Next, close the arms of the parallel lines around games $g_{122}$, $g_{322}$, $g_{155}$ and $g_{355}$. There are now four points where four lines join. Bring pairs of these points together on the extension of the diagonal formed by closing the angle between the dashed lines. Two dotted lines now form the outer boundary of the figure in the fifth panel. These lines are zipped together to form the boundary between the Prisoner's Dilemma and the anti-Prisoner's Dilemma.

Deforming the three-sided regions into equilateral triangles produces a winged octahedron. The wings are the Arctic and Antarctic regions of Figure 4.7. Figure 4.11 provides the pattern for the winged octahedron and instructions for folding it to form a three-dimensional solid.

There are other 12-game subspaces that have the same topology as the subspace of symmetric games. For each there is a corresponding winged octahedron.

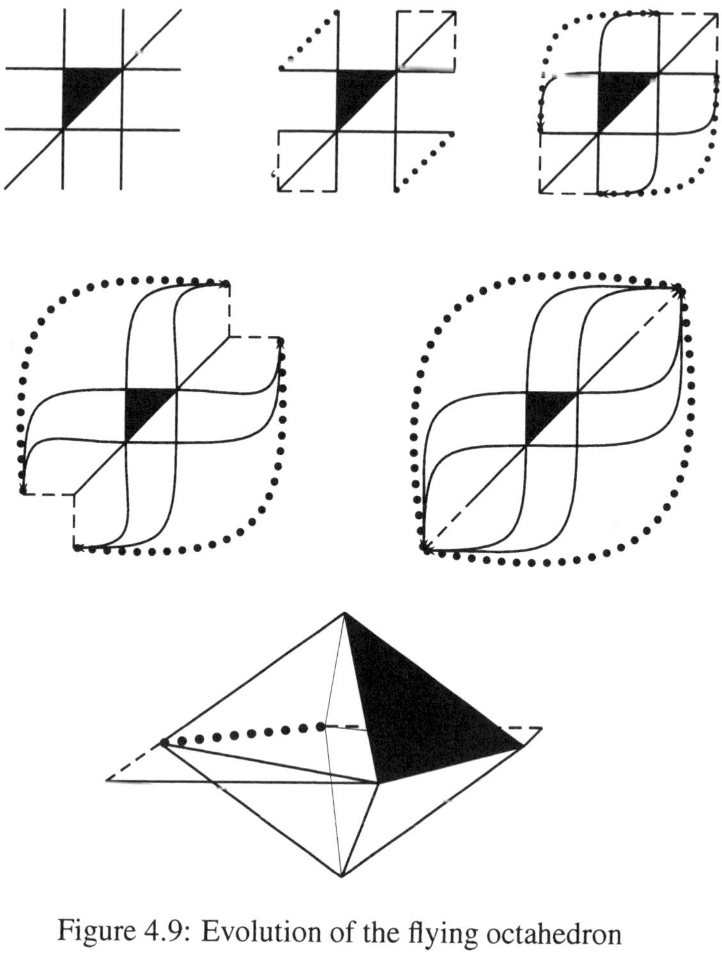

Figure 4.9: Evolution of the flying octahedron

## 4.6 Do the symmetric games matter?

Symmetry is clearly a very special case. If games were generated randomly, only 12 in 144 would be symmetric. It makes sense to ask how important the symmetric games really are.

Reasons for giving them special attention them fall into two categories. First, the symmetric games are easy to present and analyse

– they are teachable. Second, we really are interested in situations where players have equal opportunities. The symmetric games describe an egalitarian universe.

How often even ordinal symmetry appears in the real world is an open question. A great many situations are approximately symmetric, and symmetric payoffs provide a useful starting point for analysis. On the other hand, symmetric games present a problem that human beings are equipped to evade. Symmetry theoretically erases distinctions between players, but real people are capable of exploiting very subtle distinctions. For example, who goes through a door first? This is an apparently symmetric situation describable by either a Coordination game or a BoS. The problem is generally resolved by both players using conventional markers like sex or age to break the symmetry. The interesting symmetric games, with multiple or inferior equilibria, will be of considerably more practical interest if symmetry is not strictly necessary. The next chapter looks at asymmetric versions of one symmetric game.

One message of this chapter is that symmetric games provide a conveniently small model for introducing analytical techniques, including some from topology and group theory. For teaching purposes it is an advantage that most of the familiar games are symmetric.

Another message of the chapter is that symmetry relationships are important in understanding the $2 \times 2$ games. Symmetric games provide a very direct approach to them.

Third, although the symmetric games are the best known and the most studied of the 144 distinct ordinal games, a great deal that is interesting about them has been overlooked. The topological approach provides a systematic approach that brings out similarities and patterns that are not apparent otherwise.

## 4.7 Appendix: Other subspaces under the symmetric operations

The symmetric operations produce a complete and disjoint partition of the 144 $2 \times 2$ games that provides insight into the structure of the space of $2 \times 2$ games. The subspaces are aesthetically appealing, but appear to have no behavioural implications.

The seven basic payoff combinations and 17 wirings were presented as Figure 2.9. Games based on each pattern appear only in specific subspaces generated by symmetric operations. The symmetries in the entire space of 144 games can be described in terms of the distribution of basic wirings and the subspaces generated with symmetric operations.

Figure 4.10 shows the locations of the subspaces generated by the symmetric operations within the standard layout. In every case the subspace lies on a diagonal of the four piled layers. There are subspaces of 6, 12 and 24 games, each with its own pattern of links.

### Six-game subspaces

The 6-game subspaces are simply the cycles of games on the negative main diagonal of each layer. These are the quasi-symmetric games, shown in Panel 3 of Figure 4.10 for a typical layer.

An interesting feature is the alternation of single and double links. The negative diagonal connects tiles for which $R_{34}$ and $C_{34}$ lead to the same level. As a result, the symmetric operation, $S_{34}$, returns to the same layers, producing the same result as $S_{12}$.

On Layers 1 and 3, the three games on the left are R&G reflections of the three games on the right. Games on the left of Layer 2 are R&G reflections of the games on the right on Layer 4.

### Twelve-game subspaces

The 12-game subspaces all resemble the symmetric games in the pattern of linkages and in certain features of the games themselves. Panel 2 in Figure 4.10 shows in black and grey the locations of four 12-game subspaces that are not on the main diagonal. The sets of

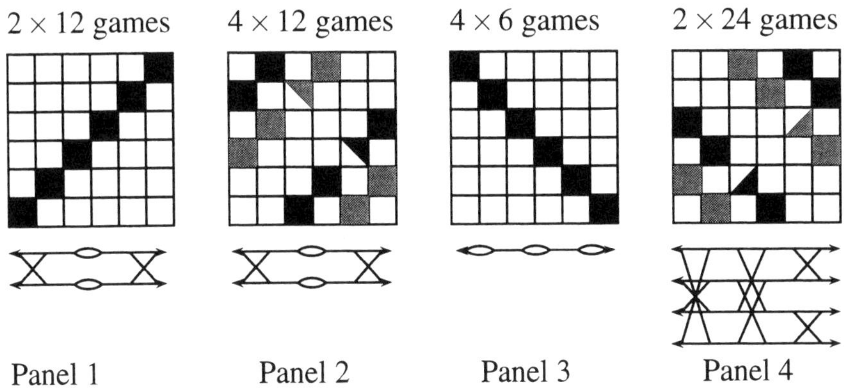


Figure 4.10: Symmetric structures

games indicated in grey are R&G reflections of those in black. There is also one other 12-game subspace that shares the main diagonal with the symmetric games in Panel 1.

**Twenty-four-game subspaces**

Parallel to the main negative diagonal, there are two subsets of 24 games marked in black and gray in Panel 4 of Figure 4.10. Both groups contain twelve pairs of R&G reflections. The 48 games in these two subspaces are transformations of the three wirings on the one completely asymmetric payoff pattern shown in Figures 2.9.

The games in each group of 24 are linked the same way as the slices discussed in Figure 3.4. There are no double links.

| *number of swaps* | *swap type* | *number of versions* | *number of games* | *number of sub-spaces* | *description* | *group* |
|---|---|---|---|---|---|---|
| 3 | symmetric | 1 | 6 | 4 | | |
| | | | 12 | 6 | | |
| | | | 24 | 2 | S-slice | $P_3$ |

Table 4.1: Subspaces with symmetric swaps

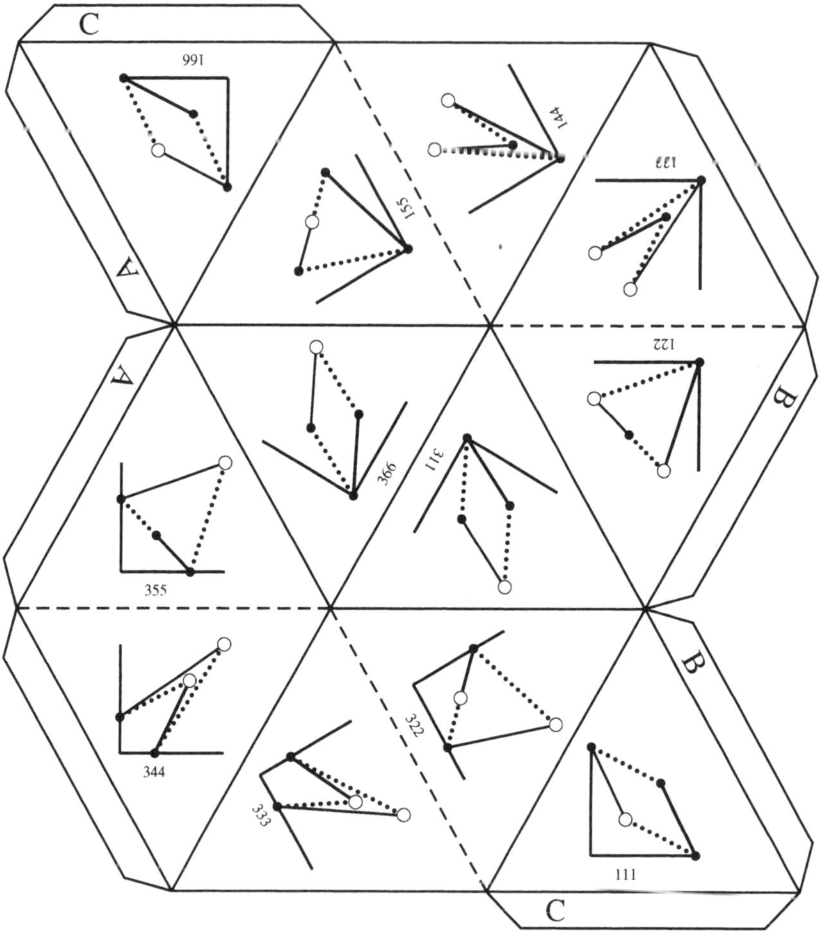


1. Fold back and crease all edges including tabs
2. Fold dashed edges forward
3. Connect tabs A
4. Connect tabs B
5. Connect tabs C

Figure 4.11: Winged octahedron

# Chapter 5

# A Family for the PD

## 5.1 The most famous game

The Prisoner's Dilemma is the most famous $2 \times 2$ game of all. Since its introduction by Albert W. Tucker in 1950, it has become the workhorse of introductory game theory. Robert Axelrod called it the *e. coli* of the social sciences. McCain even suggested that the Prisoner's Dilemma was as important as von Neumann and Morgenstern's *Theory of Games and Economic Behavior*. Jon Elster has called the generalized Prisoner's Dilemma the fundamental problem in political science.

The Prisoner's Dilemma matters to theorists because it provides a vivid and widely known objection to the First Theorem of Welfare Economics, the most important theorem in the social sciences. The First Theorem states the conjecture embedded in Adam Smith's famous "invisible hand" metaphor that, under some conditions, independent rational choice will lead to a "good" allocation. For the theorem, "good" is interpreted as Pareto-efficient.

The Prisoner's Dilemma demonstrates the existence of cases in which independent rational choice leads to a Pareto-inefficient outcome. The result pivots on the presence of reciprocal negative externalities that the First Theorem rules out: any time a player is in a position to improve her payoff by one rank, the move will worsen the payoff for the other player by two ranks. The Prisoner's Dilemma is therefore a complement to the First Theorem rather than a counterexample.

The Prisoner's Dilemma, or PD, is only one of 144 strict ordinal games, but, as we showed in Chapter 3, it lies in a connected region containing other games with unique Pareto-inferior equilibria. The existence of games closely resembling the Prisoner's Dilemma suggests that the problems it illustrates may be even more common than they appear.

In this chapter we explore a family of games which we call the Prisoner's Dilemma Family or PDF. Introducing the Prisoner's Dilemma Family requires that we identify the essential features of the Prisoner's Dilemma. Knowing the features of the $2 \times 2$ games that identify a member of the Prisoner's Dilemma Family helps when we want to generalize the Prisoner's Dilemma to larger games.

We go on to examine the location of the PDF in the overall structure of $2 \times 2$ games. We show that the Prisoner's Dilemma actually occupies a central position among the $2 \times 2$ games, lying at a kind of crossroads in the topological space.

## 5.2 The nature of the Prisoner's Dilemma

The Prisoner's Dilemma is the minimal game that demonstrates the Hobbesian dilemma in which actions that are individually beneficial are socially harmful. Most real-world situations described as PDs – arms races, common property problems, free-rider problems, public goods – are actually multi-person or multi-stage games. There is also a broader literature on "Social Dilemmas", a class of problems which includes the PD as well as Coordination games. What is lacking is a coherent notion of how to generalize the PD so that its relationship with other problematic games is precisely described and so that the social and economic implications are clearly connected to specific features of the payoff function.

The Prisoner's Dilemma has three notable features: it is a rank-symmetric game (*condition **S***); both players have dominant strategies (*condition **2DS***); and the outcome is Pareto-inefficient (*condition **PI***). The presence of a Pareto-dominated equilibrium is the most interesting of these for economists and other social theorists.

Symmetry and the strong equilibrium concept make the Prisoner's Dilemma a compelling, elegant and teachable example. Rank symmetry was a feature of the "non-cooperative pair" in the experiment conducted by Merrill Flood and Melvin Dresher in January of 1950 and first described in print by Flood in a RAND memorandum in 1952 [9]. The version named the Prisoner's Dilemma by Albert Tucker in the lecture to Stanford University's department of Psychology in May of 1950 was cardinally symmetric [36].

Symmetry, however, is not necessary to uniquely select Prisoner's Dilemma from among the 144 $2 \times 2$ games. Only the Prisoner's Dilemma has a Pareto-inferior dominant-strategy equilibrium. Nor is the presence of a dominant strategy equilibrium necessary. We can replace it with the weaker requirement that the game be dominance-solvable[1] (*condition **1DS***) if we retain symmetry. We can also abandon dominance altogether, describing the Prisoner's Dilemma as a symmetric game with a Nash equilibrium (*condition **N***) that is Pareto-dominated, providing that we include a condition that excludes games with multiple equilibria. *Condition **U*** is the requirement that the Nash equilibrium be unique.

### Three descriptions of the PD

We now have three descriptions with successively weaker equilibrium concepts that select the PD and only the PD from among the $2 \times 2$ games[2].

$$2DS + PI \Rightarrow PD \tag{5.1}$$

$$1DS + S + PI \Rightarrow PD \tag{5.2}$$

$$N + U + S + PI \Rightarrow PD \tag{5.3}$$

The three equilibrium concepts are nested: $2DS \subseteq 1DS \subseteq N$. Weakening the equilibrium concept requires additional restrictions. Symmetry serves in the second definition while symmetry and uniqueness are necessary in the third to ensure that a single game is chosen.

---

[1] A game is dominance solvable if a unique outcome can be found by successively eliminating dominated strategies.

[2] There is a fourth characterization, which we introduce below, that imposes conditions on the inducement correspondences and provides a natural link to larger games. In this approach the Prisoner's Dilemma is described in terms of externalities.

For the $2 \times 2$ games, the three definitions, 5.1 – 5.3, yield the same game. For $3 \times 3$ games, the set of symmetric dominance solvable games with inferior equilibria is larger than the set of symmetric games with inferior dominant strategy equilibria. Descriptions that are equivalent for $2 \times 2$ games are not equivalent for larger games. Which description then yields the appropriate $3 \times 3$ analogue of the Prisoner's Dilemma?

## 5.3 Overlapping neighbourhoods

To this point we have not considered whether the topology of the $2 \times 2$ games throws any light on the essential nature of the Prisoner's Dilemma. In fact it does.

The conditions ***2DS***, ***1DS***, ***N***, ***U***, ***S*** and ***PI*** define subsets of the $2 \times 2$ games. For each definition, the intersection of the subsets contains the single game $g_{111}$, the Prisoner's Dilemma. In the topological space, each condition corresponds to a connected region. The Prisoner's Dilemma lies in the intersections of these regions.

### The dominant strategy layout

The standard layout of a layer highlights the pattern of linkages between tiles and layers. Descriptions of the Prisoner's Dilemma involve dominant strategies, however. It therefore helps to shift the frame of reference to emphasize the continuity of the regions with dominant strategies.

Figure 5.1 shows Layer 1 rearranged so that the games with dom-

Figure 5.1: The dominant strategy layout for Layer 1, showing position of $g_{111}$, Prisoner's Dilemma

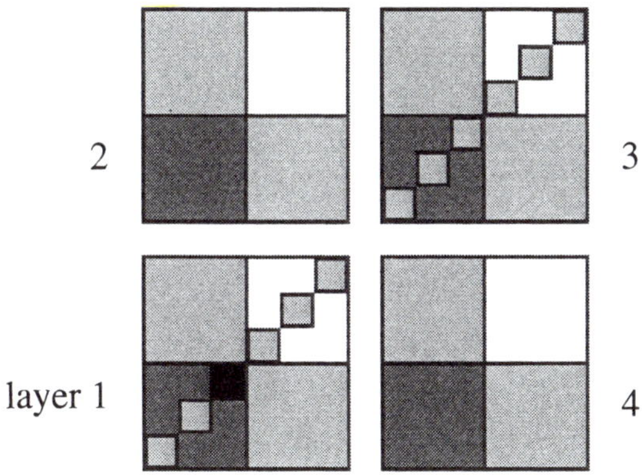


Figure 5.2: Four layers in dominant strategy layout – symmetric games on the diagonal

inant strategies for Row are in the bottom three rows, and games with dominant strategies for Column are in the left three columns[3]. Games with a dominant strategy for *both* players are in the lower left dark grey quadrants. Games for which only one player has a dominant strategy are light grey.

This *dominant strategy layout* emphasizes dominance solvability[4]. The location of the Prisoner's Dilemma in the dominant strategy layout is shown with a black square. It sits at the crossroads where the three regions meet, diagonally opposite the region where neither player has a dominant strategy. This region, shown in white, holds many of the most interesting games of all.

## Symmetry

Symmetry sometimes plays a role in defining the Prisoner's Dilemma. Symmetric games appear only on Layers 1 and 3. Figure 5.2 shows

[3]Since the payoff pattern for Row is constant in each row, Row has a dominant strategy for every game in the row, or for none. Half the games have dominant strategies, so Row has a dominant strategy in three complete rows, 1, 5 and 6 in the standard layout. Columns 1, 5 and 6 have a dominant strategy for Column.

[4]In the 1966 typology developed by Rapoport and Guyer [23] the first level of classification was based on whether zero, one or two players have a dominant strategy.

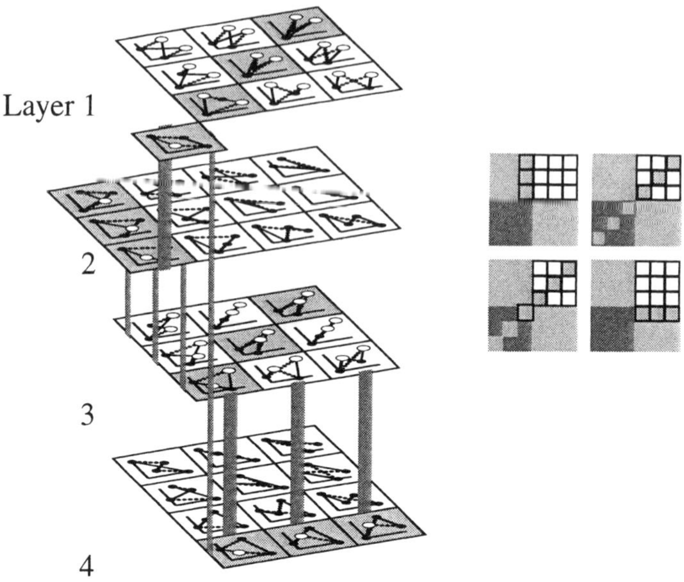


Figure 5.3: Corner of 4 layers with some $X_{34}$ links – PD and PDF

all four layers in dominant strategy layout with the symmetric games shown in light grey on the main diagonal of the combined figure. Now we see that the Prisoner's Dilemma, in black, is at the intersection of regions containing games with dominant strategies and the band containing the symmetric games.

### Linking games with Pareto-inferior equilibria

The games with Pareto-inefficient Nash equilibria do not appear on a single layer in the standard layout. The coordination games are in the white region on Layer 3. Small groups of asymmetric games with inefficient equilibria appear on Layers 2 and 4. The games are connected by $X_{34}$ swaps however.

Figure 5.3 is an exploded view of the region north-east of the Prisoner's Dilemma. Nine games with no dominant strategy are shown on each layer. Symmetric games are in grey on the diagonals of Layers 1 and 3. The Prisoner's Dilemma is attached to the corner of the block on Layer 1. Groups of games with Pareto-inefficient dominant strategy solutions are attached to Layers 2 and 4.

Some key connections between games on different layers are also shown. A vertical solid line ($R_{34}$ swap) connects the Prisoner's Dilemma to a band of three games on Layer 2 that are, in turn, connected by $C_{34}$ swaps to Coordination games on Layer 3. Another $C_{34}$ swap connects the Prisoner's Dilemma to a similar band of games on Layer 4.

### Restitching the dominant strategy layout

Figure 5.3 emphasizes links that are hard to see in layer-based configurations. In particular, the Prisoner's Dilemma is shown as a neighbour of two sets of games with Pareto-dominated Nash equilibria on Layers 2 and 4. These groups are connected to nine games on Layer 3 with Pareto-dominated Nash equilibria . The games with Pareto-dominated Nash equilibria therefore form a connected region in the topological space defined by the swap operators.

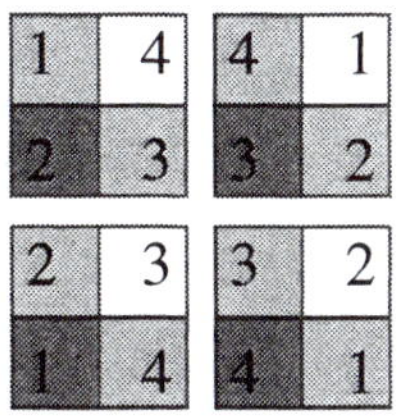


Figure 5.4: Restitched dominant strategy layout with layer of origin identified

To make these connections more obvious, we rearrange the dominant strategy layout. The edges between quadrants in Figure 5.2 represent $X_{12}$ swaps on the layers and $X_{34}$ swaps *between* layers.

The layers can be split at the $X_{12}$ edges between the quadrants and the quadrants can be restitched using $X_{34}$ swaps. Figure 5.4 shows the rearranged layout with a label in each quadrant showing the layer it came from.

In the new arrangement, all the games with Pareto-dominated Nash equilibria are on a single layer. Figure 5.5 shows an exploded view like Figure 5.3, but now all the games that interest us are on

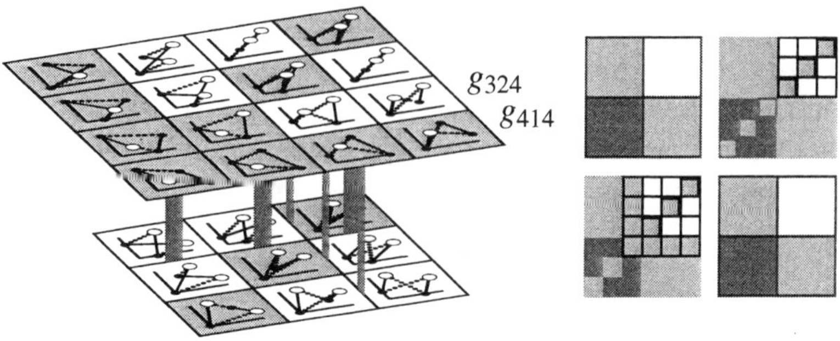


Figure 5.5: Two restitched layers showing PDF and social dilemmas

just two layers. All the $X_{34}$ swaps between layers in Figure 5.3 are now within-layer swaps.

Figure 5.5 was developed to display the games related to Prisoner's Dilemma, but the resulting diagram includes all the games in which rational individual choice can yield perverse results. Additional $X_{34}$ swaps have been introduced to show important connections between games on (original) Layers 1 and 3. The apparently disparate collection of games is actually a region in the topological space, tightly connected by swap operations.

### 5.3.1 Conditions defining the PD as intersecting regions

Figure 5.6 shows how the three definitions, 5.1 – 5.3, of the Prisoner's Dilemma appear on this restitched dominant strategy layout. In the diagram on the left, dominant strategy equilibrium (***2DS***) and Pareto-inferiority (***PI***) uniquely identify the Prisoner's Dilemma.

In the middle diagram, the weaker criterion, dominance solvability (***1DS***), intersects with ***PI*** in seven games. The symmetry condition (***S***) must be added to isolate Prisoner's Dilemma. ***PI*** and the weakest equilibrium condition, existence of a Nash equilibrium (***N***), must be supplemented with the uniqueness criterion (***U***) *and* symmetry (***S***) to identify Prisoner's Dilemma alone in the third diagram.

In the centre and right diagrams, symmetry is required to iden-

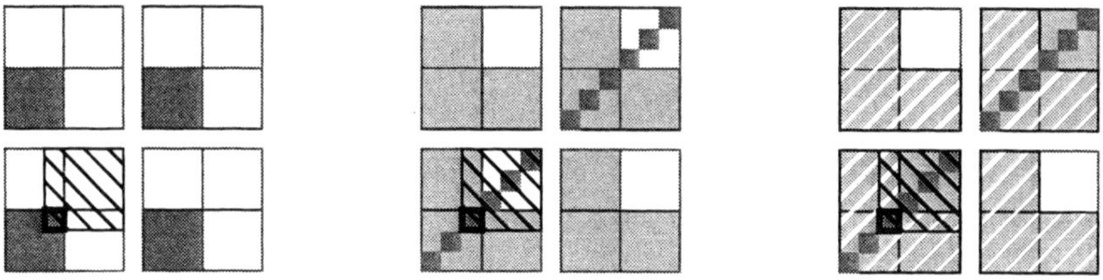


Figure 5.6: The conditions defining the PD

tify uniquely the Prisoner's Dilemma. Without this constraint, the remaining conditions intersect in an L-shaped region containing 7 games, each with a single, Pareto-dominated equilibrium. We call this set of games (outlined in Figure 5.7) the Prisoner's Dilemma Family, or PDF games. More formally,

$$1DS + PI \Rightarrow PDF \tag{5.4}$$

$$N + U + PI \Rightarrow PDF \tag{5.5}$$

Two conditions normally associated with the Prisoner's Dilemma have been dropped in defining the Prisoner's Dilemma Family: (i) PDF games need not have dominant strategy equilibria (though they are dominance solvable), and (ii) they need not be symmetric. The Prisoner's Dilemma is the only symmetric PDF game, and the only one with a dominant strategy equilibrium.

## 5.4 The Prisoner's Dilemma Family

The PDF games are interesting for the same reason the Prisoner's Dilemma is interesting – they represent situations in which rational

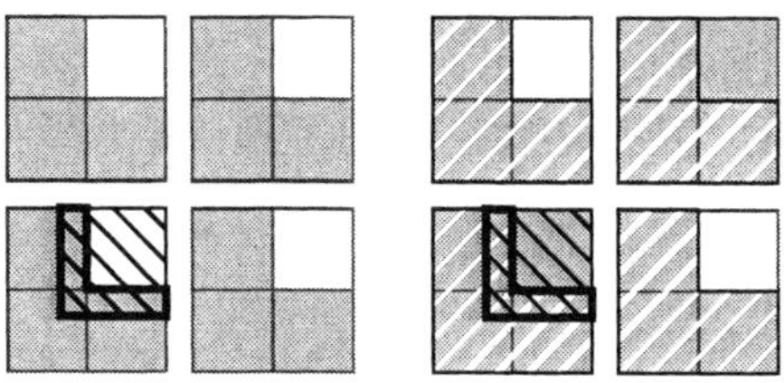


Figure 5.7: The conditions defining the PDF

individual choice leads to sub-optimal outcomes.

The Prisoner's Dilemma and the Coordination games share a feature that has made them the subject of much attention: both have Pareto-dominated equilibria. The PDF games share the feature.

## Social dilemmas

Any group of games that is analytically important and topologically connected warrants a name. The term "Social Dilemmas" has been applied to variants of the Prisoner's Dilemma, to Coordination games, to the Stag Hunt, and sometimes to the Battle of the Sexes. A strong case can be made on both topological and conceptual grounds for calling the entire 25 game region in Figure 5.5 *Social Dilemmas*. In this view, a Social Dilemma is a game in which individual rational choice does not reliably lead to socially optimal outcomes[5].

It might be argued that the Battle of the Sexes games should be excluded because the problem in these games is that distributional conflict may make achieving any Nash equilibrium difficult, not that there is an inferior Nash equilibrium[6]. The counter-argument is that if the term Social Dilemma is not to be identified with a specific game there will necessarily be different types of social dilemma.

It is significant that the Prisoner's Dilemma Family and the Coordination games are separated by $X_{34}$ swaps. Swapping the highest payoffs generally changes a game significantly, which is why we organize the games in layers distinguished by $X_{34}$ swaps. In this case the $X_{34}$ swap separates games which are similar in having a Pareto-dominated equilibrium, but differ in whether the dominating point is stable. Compare, for example, PDF game $g_{414}$ and its neighbour, Coordination game $g_{324}$ in Figure 5.5.

Although the Prisoner's Dilemma seems to be a kind of freak

[5]Eaton [8] has shown that a very wide range of models from various fields in and related to economics can be seen as variations on a basic continuous strategy space model of the Prisoner's Dilemma. He calls this collection of PD models social dilemmas. Rather than restricting the notion of Social Dilemma it might make more sense to expand the notion of a Prisoner's Dilemma.

[6]If an equilibrium in mixed strategies were defined for these games, for instance if we move from the ordinal game to the real-valued representative game, the mixed strategy equilibrium is also Pareto- dominated by either of the Nash equilibria.

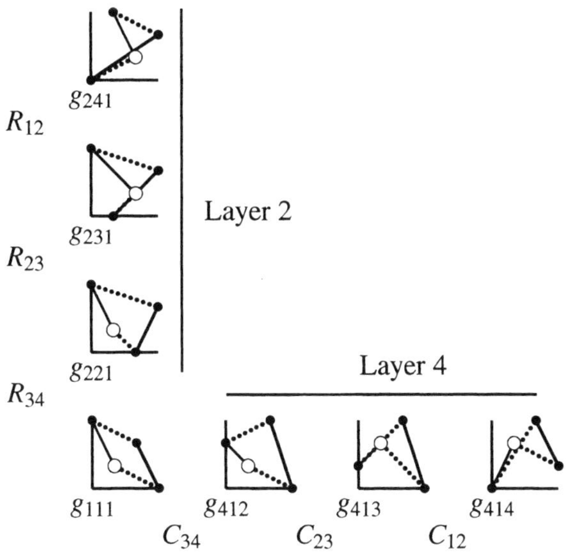


Figure 5.8: The Prisoner's Dilemma Family

among games, Figure 5.5 shows that it actually occupies a central position at the point where regions with zero, one and two dominant strategies meet. With other members of the PDF family, the Prisoner's Dilemma forms a hinge connecting the most interesting $2 \times 2$ games.

## 5.5 An alibi for one prisoner

The Prisoner's Dilemma takes its name from a story invented by A. Tucker. A curious payoff pattern had been discovered earlier the same year by Dresher and Flood. Tucker set out to make the pattern accessible and entertaining for members of the Psychology Department at Stanford University. Tuckers's story explains how the payoffs might arise, dramatizing an otherwise abstract set of values.

The players in Tucker's story are two prisoners suspected of a crime. The payoffs are actually sentences chosen by the prosecutor to give each prisoner an incentive to confess. Ordinal equivalents are given in the table below. The payoffs in fact give each prisoner an incentive to confess whether or not the other player confesses. In other words, the game is designed to give each player a dominant strategy.

Naturally the prosecutor arranges the payoffs so that both prisoners receive longer sentences if they both confess. The outcome is thus Pareto-inferior, at least from the point of view of the prisoners. The incentive structure is symmetric because the same story is told to both prisoners.

Tucker's story can be reworked to describe asymmetric games. Say one player, Column, has a watertight alibi for the minor crime. This amounts to swapping payoff 3 for payoff 4 for Column in the Prisoner's Dilemma. The result is $g_{412}$. The prosecutor can still tempt Row to confess by offering a reduced sentence in exchange for a confession: confessing is still a dominant strategy for Row. The rest of the standard story applies. The prosecutor tells Column that if Row confesses and Column does not, Column will receive a long sentence. If both confess, Column will get a shorter term, the same as Row's.

The game still presents the players with a dilemma. Column can see that Row has a dominant strategy. Knowing that, Column should expect Row to confess. If Row confesses, Column is better off confessing as well. Row also needs to consider Column's potential behaviour. Column does not have a dominant strategy as in the PD, but Row knows that Column knows that Row has a dominant strategy. Row understands that Column would be unreasonable to expect

---

## The Classic Prisoner's Dilemma

Two men, charged with a joint violation of the law, are held separately by the police. Each is told that

| | | payoffs |
|---|---|---|
| 1 | if neither confesses, both will receive a short sentence | [3, 3] |
| 2 | if one confesses and the other does not, the former will be | [1, 4] |
| | set free and the latter will be given a long sentence | [4, 1] |
| 3 | if both confess, each will be given a moderate sentence | [2, 2] |

Row to play a dominated strategy. Row cannot expcct Column to trust him.

The description of how the players might reason relies on *dominance solvability*. Column reasons that Row would expect her to eliminate the dominated strategy. It also relies on the assumption that each knows the other is rational.

Similar stories can be constructed for $g_{413}$ and $g_{414}$. They differ from $g_{412}$ only in that the rank-cost of being the only one to confess for the player with an alibi is relatively larger. Row might believe that Column is less likely to confess in $g_{414}$ than in $g_{412}$. Since Row's payoffs are unchanged, however, Column has at best a subtle case for thinking that Row would behave differently in $g_{414}$ than in $g_{412}$.

## 5.6 The asymmetry of the Alibi games

As far as we know, the asymmetric members of the PDF games have not been studied. It is easy to show that the games have features that are not present in the PD. For example, $g_{412}$ has a rank-symmetric equilibrium despite its asymmetric payoff matrix. Game $g_{413}$, on the other hand, has an asymmetric Nash equilibrium. It is not obvious how people will play such games. What does seem clear is that conclusions about behaviour in games with inferior Nash equilibria should not be based solely on the symmetric Prisoner's Dilemma.

### 5.6.1 Evolution with PDF games

One way to explore behaviour in the PDF games is to conduct simulated evolutionary experiments. We have done simple simulations on PDF games that support the notion that behaviour in Alibi games cannot be inferred from behaviour in the PD. Beaufils, Delahaye and Mathieu [3] described a small round-robin tournament of the evolutionary PD game following seven typical strategies through 1000 generations. Four strategies, led by "Tit for Tat", survived. Our results for the asymmetric games were significantly different from the results for the Prisoner's Dilemma. Tit for Tat, for example, did not

### An Alibi Game

Two men, charged with a joint violation of the law, are held separately by the police. Each is told that

| | | payoffs |
|---|---|---|
| 1 | if neither confesses, both will be given a short sentence on some pretext. One of the players has an alibi, however, *which protects him from the short sentence* | [3, **4**] |
| 2a | *if the one with the alibi confesses* and the other does not, the former will be given a *token sentence* and the latter will be *given a long sentence* | [1, **3**] |
| 2b | if the one without an alibi confesses and the other does not, the former will be released and the other will be given a long sentence | [4, 1] |
| 3 | if both confess, each will be given a moderate sentence | [2, 2] |

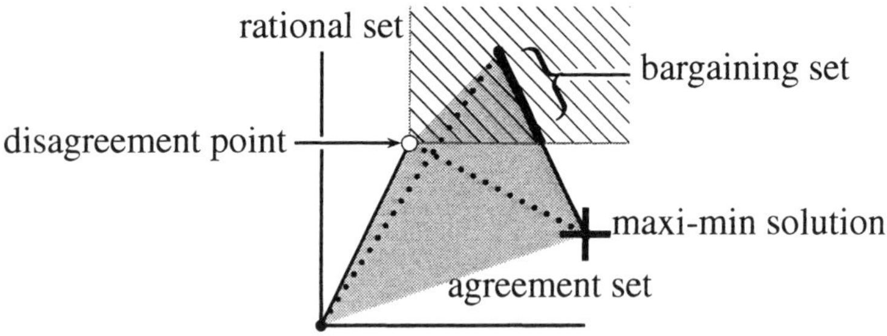


Figure 5.9: Agreement, rational and bargaining sets for $g_{414}$

survive as a strategy for either Row or Column in any Alibi game.

## 5.6.2 Bargaining in Alibi games

Another approach to behaviour in the PDF games is to imagine converting them into Nash Bargaining games. The concepts used in simple bargaining theory are illustrated with $g_{414}$ in Figure 5.9. Ordinal values are treated as rationals to illustrate the implications of asymmetries.

The main elements of a Nash Bargaining game are a *disagreement point* representing the outcome if no agreement is reached, and a convex set representing all possible imputations. For a $2 \times 2$ game, the set of outcomes that can be reached by correlated strategies in repeated games is a reasonable interpretation of the *agreement set*. It is simply the set of all possible convex combinations of the outcomes in payoff space, shown in grey.

In the case of the Prisoner's Dilemma and games $g_{221}$ $g_{231}$, $g_{412}$ and $g_{413}$, the disagreement point can plausibly be identified with the Nash equilibrium, following Binmore [4]. The disagreement point is defined as the payoffs that players receive in the absence of an agreement. This is usually interpreted as the non-cooperative Nash equilibrium.

Some authors [22] argue the disagreement point represents the payoffs that players can guarantee themselves against the malevolence of the other player, and conclude that the maxi-min outcome is the better choice for the disagreement point. The two views agree for most of the PDF games, but for $g_{241}$ and $g_{414}$ the Nash equilibrium differs from the maxi-min solution. The maxi-min solution is shown in Figure 5.9. Using the maxi-min yields a different outcome for the hypothetical bargaining process.

Nash's rationality axiom restricts the bargaining solution to the subset that Pareto-dominates the disagreement outcome, shown as the hatched region. The efficiency axiom limits outcomes to the undominated members of the agreement set. This constrains the solution to the negatively sloped portion of the boundary of the agreement set. The *bargaining set*, shown as a heavy dark line segment joining (3, 4) and (3.5, 3), is defined as the undominated and rational subset of the agreement set[7].

### Solutions

A solution to the bargaining problem is simply a rule that picks out one member of the bargaining set. For the Prisoner's Dilemma the two most popular solutions, Nash and Kalai-Smorodinski (KS), coincide. For the Alibi games the solutions differ. Figure 5.10 illustrates the two solutions (larger white circles) for $g_{412}$.

[7] Notice that if the mini-max solution is taken as the disagreement outcome it is also the bargaining outcome by the argument in this paragraph.

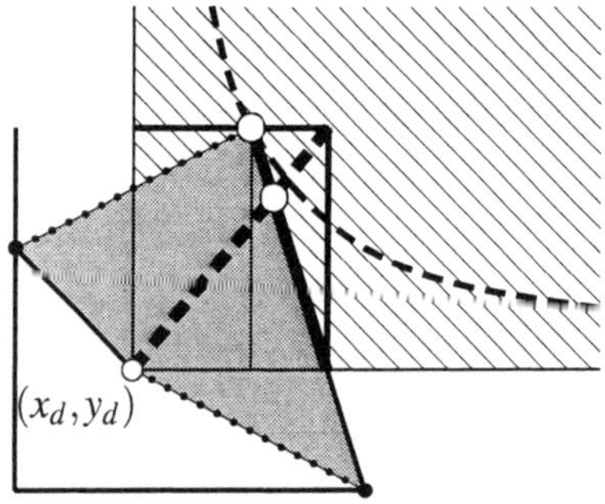


Figure 5.10: Nash and KS solutions for $g_{412}$

The Nash solution is the point in the bargaining set that maximizes the area of a rectangle with one corner at the disagreement point $(x_d, y_d)$ and the opposite corner at the solution. On the diagram it is the point where the bargaining set touches the hyperbolic curve $(x - x_d)(y - y_d) = k$ with maximum $k$.

The Kalai-Smorodinski solution differs from the Nash solution in taking into account the most optimistic outcomes available to each player: it can be found on the graph by connecting the disagreement point to the upper right corner of the smallest rectangle containing the bargaining set within sides parallel to the axes. The corner of the minimal rectangle is $(3.67, 4)$. The line connecting $(2, 2)$ and $(3.67, 4)$ cuts the bargaining set at a point closer to an equal outcome than the Nash solution.

Features of the two bargaining solutions for the PDF games are summarized in Figure 5.11. In $g_{413}$ and $g_{414}$, the Nash bargaining outcome is $(3, 4)$. At that point both players enjoy equal rank gains, although the outcome is unequal. Only in $g_{412}$ does the Nash solution yield unequal rank gains.

The key observation is that the Nash and Kalai-Smorodinski solutions differ for the asymmetric games. Like the simulations described above, treating the PDF games as bargaining games suggests caution about generalizing from research on the PD.

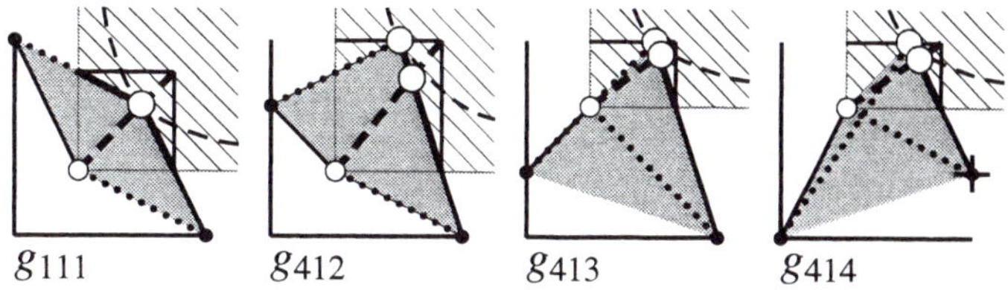


| | $g_{111}$ | $g_{412}$ | $g_{413}$ | $g_{414}$ |
|---|---|---|---|---|
| Disagreement | Equal | Equal | Unequal | Unequal |
| Nash Payoff | Equal | Unequal | Unequal | Unequal |
| Nash Gain | Equal | Unequal | Equal | Equal |
| KS Payoff | Equal | ....... More equal than Nash ....... | | |

Figure 5.11: Bargaining solutions to PDF games

## 5.7 Rank-sum inefficiency in the PDF

One of the most interesting characterizations of the PD is that it is the only game for which every best response is (rank-sum) inefficient. The *rank-sum* is analogous to the total payoff in real-valued games. We define it as the sum of the ranks of the payoffs for the players.

Rank-sum inefficiency of the best response is the ordinal equivalent of a situation in which private incentives result in such large externalities that private incentives are misleading. In the Prisoner's Dilemma, on the left in Figure 5.12, *every* individually-improving move reduces the rank-sum: individual incentives always induce externalities larger than the private gains.

In all the Alibi games, the rank-sum efficient outcome is difficult to sustain because at least one player has an individually rational

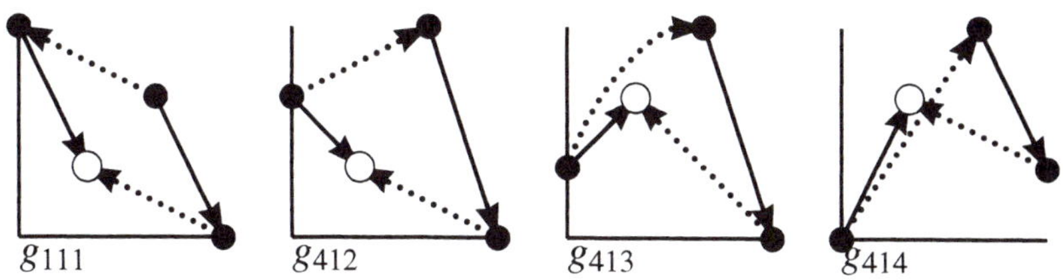


Figure 5.12: Direction of unilateral moves in the PDF games

alternative which is socially inefficient, as the arrows on the inducement correspondences show. One player always prefers to move away from the efficient point.

Furthermore, at the new position the other player has an incentive to move. In every PDF game, the second move harms the first player and, in $g_{412}$ and $g_{414}$ as well as $g_{111}$, the second move is also rank-sum inefficient. (In $g_{413}$, the second move is rank-sum neutral.) The overall effect of the sequence of individually rational moves is rank-sum inefficient[8].

Characterizing the Prisoner's Dilemma and the PDF in terms of the social inefficiency of private decisions has several advantages.

1. It is an explicitly economic characterization.

2. It identifies the aspect of the PD that is socially significant.

3. It is consistent with common usage about larger games.

4. It provides a very simple criterion for identifying "perfect" N-person Prisoner's Dilemma: a game in which every best response is socially inefficient is a perfect PD.

This characterization provides a natural way to generalize the PD to the asymmetric case and to multi-player, multi-strategy games.

## 5.8 Concluding remarks

The Prisoner's Dilemma is one of the 43 games in which rational individual action does not automatically yield a desirable outcome. It is one of 25 games with Nash equilibria in which rational individual action can lead to undesirable outcomes. We are calling these the Social Dilemmas. The PD is one of 16 games in which rational choice can even lead to a Pareto-dominated outcome. It is one of

[8]In multi-player games the externalities might be distributed over many players. This is the justification for describing public common property games, emission games, and even free-rider problems as multi-person Prisoner's Dilemmas. The sum of losses in such games still exceeds the private gains for the decision-maker.

seven, the games we have called the Prisoner's Dilemma Family, in which a unique Nash equilibrium is Pareto-dominated. In our view, this is the significant feature of the PD, the feature that identifies an economically interesting class of problems and which allows us to generalize to larger games.

The PD is compelling for two reasons: the outcome is inferior in a strong sense – it is Pareto-inefficient – and it is supported by a solution concept that is difficult to resist – it has a dominant strategy equilibrium. Symmetry, which is unique to the Prisoner's Dilemma among PDF games, is essentially an expository convenience. We have argued that the remaining members of the PDF, the Alibi games, deserve attention as they share the Pareto-inefficient outcome. They are also dominance-solvable.

We can only speculate about how often the asymmetric relatives of the PD occur in social situations. Each of the six Alibi games is as likely to appear in a randomly generated payoff matrix as the PD. What we don't know is whether the payoffs in the real world are generated randomly or whether nature has a bias in favour of symmetry. If nature is completely unbiased, only one in seven PDF games will be rank-symmetric.

It is difficult to say whether Alibi games are important until we search for real world examples. The Prisoner's Dilemma was originally seen as a curiosity. It took time to recognize that multi-person versions like the common property resource problem, the free-rider problem, and multi-period escalation games like the arms race are both common and practically important.

Since symmetric games are easier to construct and to explain, it is possible that the main reason that social situations corresponding to Alibi games have not been described is that no one has been looking for them. It is also possible that some Alibi games have been misidentified as Prisoner's Dilemmas. The apparent real-world ubiquity of the Prisoner's Dilemma may even be illusory – all those situations with nasty equilibria may seem to be Prisoner's Dilemmas because we haven't looked closely. It is certainly possible that the multi-person and multi-period social dilemmas that are seen as PD analogues may be asymmetric.

# Chapter 6

# Connecting the layers

The beauty of a topological approach to the $2 \times 2$ games is that every topological feature yields some surprising insight into the relationships among the games. Even ignoring features can be productive. This chapter and the next are based on ignoring the $X_{23}$ swaps.

The entire set of $2 \times 2$ games, as we showed in Chapter 3, is generated by a group of six swap operations and an identity operator. Proper subgroups of operators generate partitions of the space. For example, if we leave out the $X_{34}$ swaps, we get the four closed toroidal surfaces that we call layers. The set of generators for a layer is $\{C_{12}, C_{23}, R_{12}, R_{23}, \mathrm{I}\}$.

If we leave out $X_{23}$ swaps, the four-layered torus breaks up into a series of six 16-game blocks we call pipes and six 8-game blocks we call hotspots. A major goal of this chapter is to describe pipes. The structure of the pipes will reveal a fundamental feature of the topology of the $2 \times 2$ games.

## 6.1 Least among equals: the $X_{12}$ swaps

If both $X_{23}$ and $X_{34}$ swaps are removed from the generator set, leaving $\{C_{12}, R_{12}, \mathrm{I}\}$, the 144-game graph breaks up into 36 *tiles* of four games. Each layer breaks up into nine tiles, and pipes break up into piles of four tiles.

Tiles are a significant feature for structural and behavioural reasons. Structurally, the tiles can be imagined as the building blocks

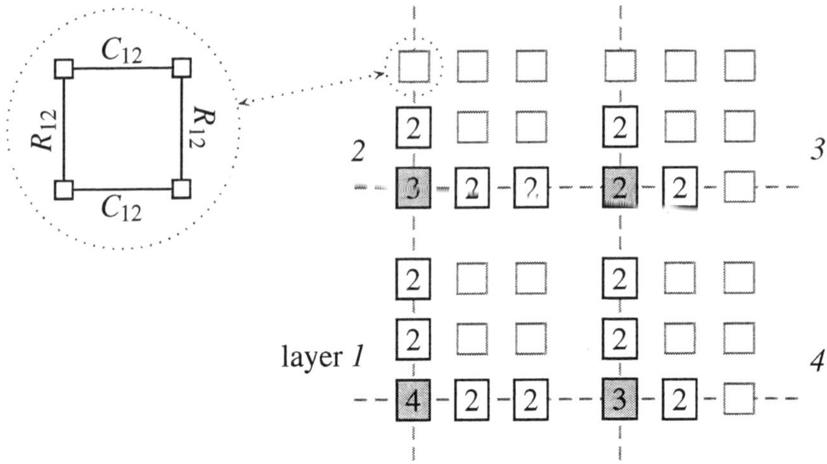


Figure 6.1: Number of equilibrium-altering $X_{12}$ swaps on each tile

of the space of the $2 \times 2$ games. $X_{34}$ links between layers weld tiles into *pipes* or *hotspots*. $X_{23}$ links join tiles to make layers.

Behaviourally, a tile associates games that differ only by $X_{12}$ swaps, which is to say, only in the ranking of the least-preferred elements. Since these are rarely selected by rational players, a majority of tiles consist of games that are economically or behaviorally equivalent. Tiles are therefore natural groups for analysis, and tiles that contain games with distinct outcomes are of particular interest.

There are 36 tiles on four layers. On 20 tiles the four games have the same equilibrium outcome. Figure 6.1 shows that the 16 tiles where $X_{12}$ swaps do affect the equilibria are located in the left columns or bottom rows of the standard layout. The numbers on the tiles indicate how many of the four swaps change the equilibrium outcome[1].

The tiles where change occurs are clustered at the bottom and left of each layer. One third of all the changes induced by $X_{12}$ swaps occur in just four (grey) tiles. The tiles here are linked by $X_{34}$ swaps to form the most heterogeneous pipe of all. It includes the Prisoner's Dilemma and Chicken. Together the 16 games in this pipe are a

[1]There are 72 $R_{12}$ and 72 $C_{12}$ swaps. Eighteen of each cause an equilibrium change between games. In contrast, $R_{34}$ and $C_{34}$ each alter the equilibrium in 60 of 72 swaps.

| | | *L* | *R* |
|---|---|---|---|
| Row 2 | *No DS for Row* | $\boxed{2}, *$ | $3, *$ |
| | | $1, *$ | $\boxed{4}, *$ |

| | | *L* | *R* |
|---|---|---|---|
| Row 1 | *Dominant Strategy for Row* | $1, *$ | $3, *$ |
| | | $\boxed{2}, *$ | $\boxed{4}, *$ |

Table 6.1: Row's payoffs when Row has a dominant strategy and $R_{12}$ may affect equilibrium

microcosm of the entire space of $2 \times 2$ games. We will use it as an example throughout this chapter.

## 6.2 Instability zone – $X_{12}$ swaps matter

Swapping the lowest-ranked payoffs for, say, Row can only affect the equilibrium outcome when Row's two lowest payoffs appear in the same inducement correspondence (i.e., in the same column of the payoff matrix). The $R_{12}$ swap then changes Row's best response in that inducement correspondence.

If Row had a dominant strategy initially, then $R_{12}$ will eliminate it; if Row does not have a dominant strategy, $R_{12}$ will create one. It follows that $R_{12}$ can only affect equilibrium behaviour at the boundary between regions with a dominant strategy and regions without a dominant strategy. Row has a dominant strategy in the bottom row of each layer (standard layout) but not in the second row.

Every one of Row's payoffs in the 24 games of the first rows on four layers exhibit the dominant strategy pattern in the lower matrix of Table 6.1. All games in the second rows have the payoff pattern of the upper matrix. Whether the swap affects the outcome depends on the pattern of Column's best responses

Column has only four patterns of best responses, of which three involve equilibrium change and one does not. Hence, 18 of the 24 $R_{12}$ swaps across this boundary change the equilibrium and six do

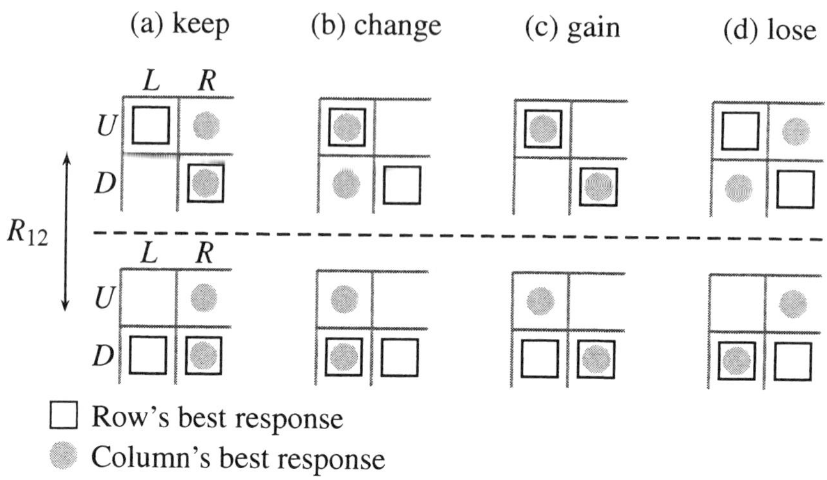


Figure 6.2: Best responses in the instability zone; only case (a) is invariant under $R_{12}$

not. Figure 6.2 illustrates the patterns in payoff matrix format. The 1 and 2 payoffs for Row are always in the left column exactly as in Table 6.1. Best response payoffs are indicated for Row with squares and for Column with circles. Concentric squares and circles identify equilibrium outcomes. Each pattern represents six games.

The pair of matrices in Figure 6.2(a) show what happens when Column has a dominant strategy ($R$) that does not select Row's inducement correspondence containing 1 and 2 . In the bottom matrix, Row has a dominant strategy ($D$) which is lost in the $R_{12}$ swap. The game remains dominance solvable however, and the equilibrium outcome is unchanged.

If $L$ is a dominant strategy for Column as in (b), $R_{12}$ causes the unique equilibrium to shift to the upper left cell.

In the matrices of (c) and (d), Column does not have a dominant strategy. Because Row has a dominant strategy in the bottom matrices, the games are dominance solvable. If the outcome with Row's 2 is not an equilibrium as in (c), $R_{12}$ creates a second Nash equilibrium. This happens on Layers 1 and 3. If the outcome with Row's 2 *is* an equilibrium as in (d), $R_{12}$ destroys it, producing a game with

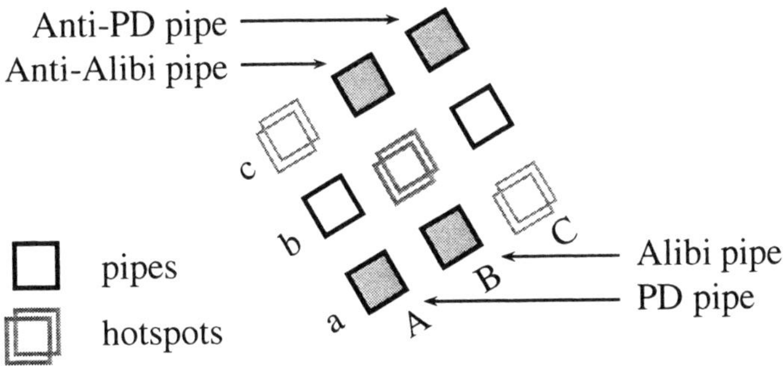


Figure 6.3: Pipes and hotspots

no Nash equilibrium. This happens on Layers 2 and 4[2].

The instability zone includes 16 tiles that contain the 64 most fragile $2 \times 2$ games . They are fragile in the sense that the smallest of changes in payoffs produce qualitatively different games.

## 6.3 Pipes at last

If we add $R_{34}$ and $C_{34}$ to the set of generators that produce the tiles, we get subspaces that appear as piles of tiles linked by $X_{34}$ swaps. The subspaces come in two forms. If $R_{34}$ and $C_{34}$ link to the same layer, tiles are connected in pairs. These we have dubbed *hotspots* and they are the topic of the next chapter. If the $X_{34}$ swaps link to different layers, all four tiles are connected. We call these 16-game subspaces *pipes*. In Figure 6.3, the hotspots are outlined in grey and the pipes in black on a standard layout. The term "pipe" correctly suggests a tube connecting the layers even though the topological structure is a bit more complicated.

[2]The reader may recall that the PDF games in Row 1 appear on Layer 4, and, by the argument presented here, must border a region with no equilibria.

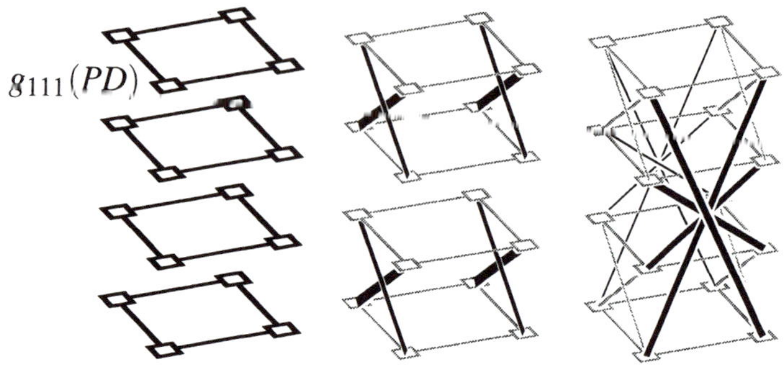


(a) (b) (c)

Figure 6.4: The PD pipe, microcosm of the $2 \times 2$ games

There are six pipes. Two pairs are R&G reflections of each other – in Figure 6.3, *Ab* reflects into *Ba* and *Cb* reflects into *Bc* – so there are only four distinct pipes to be described in the following subsections. We focus on the pipes with greyed interiors.

### 6.3.1 The pipe with the PD, microcosm of the 2×2 games

The most interesting pipe, *Aa* in Figure 6.3, is a kind of *microcosm* of the $2 \times 2$ games. It contains the Prisoner's Dilemma ($g_{111}$), Chicken ($g_{122}$), two Alibi games ($g_{121}$ and $g_{412}$), two games with two equilibria ($g_{122}$, $g_{322}$), and two with no equilibrium ($g_{222}$, $g_{422}$). Four of the games are symmetric and six pairs are R&G reflections. The pipe contains the tiles shaded grey in Figure 6.1 at the intersection of the instability zone rows and columns.

Figure 6.4(a) shows the tiles piled but not linked. Panel (b) shows the $R_{34}$ swaps. Layers 1 and 2 are connected as are Layers 3 and 4. From the Prisoner's Dilemma for example, the swap leads to $g_{221}$, an Alibi game. Since Column has a dominant strategy, the equilibrium strategy combination does not change. Panel (c) adds the $C_{34}$ swaps that link Layers 1 to 4 and 2 to 3. The pipe is now a connected graph.

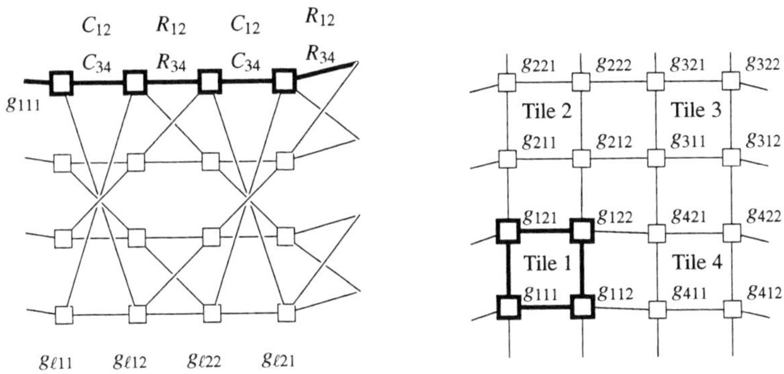


Figure 6.5: The PD pipe (a) unrolled, (b) planar

Because the microcosm straddles the symmetry diagonal of the layers, the $C_{34}$ swaps reflect the $R_{34}$. For example, the $C_{34}$ link from Prisoner's Dilemma leads to $g_{412}$, the Alibi game that is the R&G reflection of $g_{221}$.

The complete pattern of linkages in the PD pipe is shown again in Figure 6.5(a). The pipe is cut through the $R_{12}$ links in the left column and unrolled.

Figure 6.5(a) reveals a regular pattern, but one that is not immediately meaningful. It is, however, equivalent to the planar graph in (b) where the tiles appear as closed loops arranged so that the $X_{34}$ links never cross. The tile on layer one is indicated by heavy lines in both panels of Figure 6.5.

As with layers, the planar graph representing a pipe can be embedded in a torus. The Euler number

$$\begin{aligned} \chi &= V+F-E \\ &= 16+16-32 \\ &= 0 \end{aligned}$$

confirms the identification of a pipe as a torus.

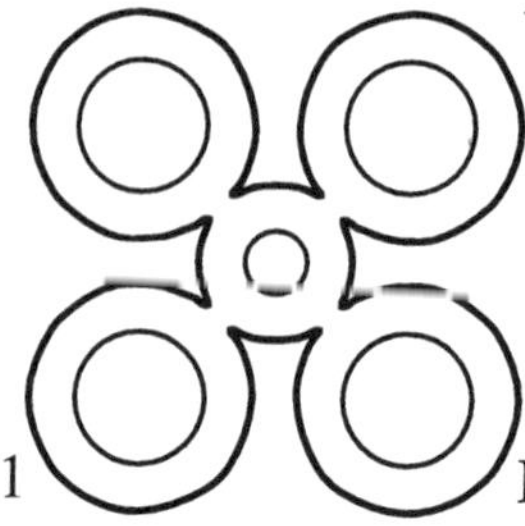


Figure 6.6: How a pipe joins the four layers

### 6.3.2 Pipes and layers

Since there are six pipes, we now have six new 16-game toruses. To understand how they relate to the four-layered torus described in Chapter 3, notice that each tile in a pipe is also a tile in a layer. [3]

Starting with a torus with the graph of a layer embedded in it and a second torus with the the graph of a pipe, imagine gluing the two toruses together so that the four points of a shared tile coincide. Now imagine puncturing the two tiles that are glued together to make a door *out* of the pipe-torus and *into* the layer-torus. We now have a figure-eight, a two-holed torus.

Since each of the four tiles in a pipe connects to a different layer a single pipe links all four layers. Figure 6.6 shows the result, a five-holed torus.

To calculate the Euler number, note that each of the five toruses has an Euler number of zero. When we join two at a shared face we have to remove two faces. We also have to remove four vertices and four edges because there are duplicates on the pipe and layers when they are separate. The adjustment when joining two toruses at a tile is therefore

$$\Delta\chi = \Delta V + \Delta F - \Delta E = -4 + (-2) - (-4) = -2$$

[3] There are other four-sided faces in both, but we are reserving the term *tile* for subspaces generated by the group $\{R_{12}, C_{12}, I\}$.

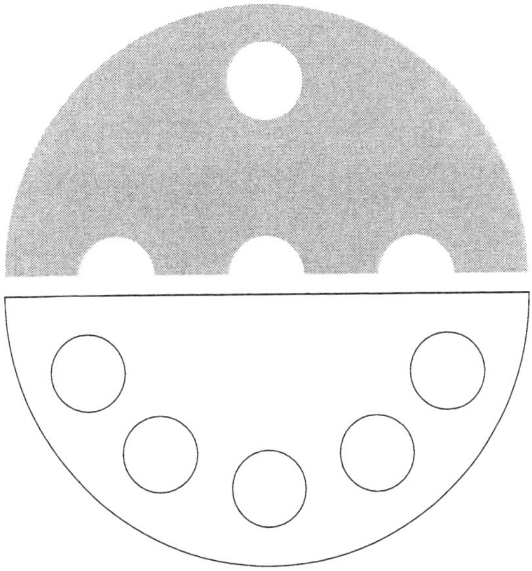

Figure 6.7: Adding a pipe

yielding as $0+0-2=-2$, the Euler number for a torus with two holes. Each layer adds another hole and reduces the Euler number by 2, so the five-holed structure in Figure 6.6 has an Euler number of $-8$.

If one connecting pipe creates a five-holed torus, what happens when we add a second pipe? Each additional pipe-torus must intersect each of the layers at one tile, so the new torus must touch the five-holed torus at four tiles. As Figure 6.7 shows, adding a second pipe adds *four* additional holes. The Euler number of the resulting structure is $(-8)+(-8)=-16$.

The same argument applies to the remaining four pipes. In other words, the first pipe connects the layer toruses into a surface with five holes while the remaining five pipes each add four holes. When all six are incorporated, there must be $5+5\times4=25$ holes and an Euler number of $(-8)+5\times(-8)=-48$ for the resulting object. All that remains is attachment of the six hotspots. In the next chapter, on page 108, we will complete the description of the topological surface of the $2\times2$ games.

| | PD pipe | Alibi | Anti-Alibi | Anti-PD |
|---|---|---|---|---|
| $R_{12}$,$C_{12}$ | $6+6=12$ | $8+0=8$ | 0 | 0 |
| $R_{34}$,$C_{34}$ | $6+6=12$ | $8+6=14$ | $8+8=16$ | $8+8=16$ |
| Total | 24 | 22 | 16 | 16 |

Table 6.2: Equilibrium changes caused by swap operations in pipes

## 6.4 Four kinds of pipes

There are six pipes. Although they are identical topologically, there are four distinct types in terms of the games that comprise them. The PD pipe has already been described. We now consider the other three, beginning with the one adjacent to it, *Ba* in Figure 6.3, which we shall refer to as the Alibi pipe.

This pipe contains two Alibi games, four with no equilibrium and four with two equilibria. The remaining six games are dominance-solvable with efficient equilibrium outcomes. Obviously, the equilibrium conditions in the pipe vary considerably. Like the PD pipe, it crosses the boundary between games that do or do not have a dominant strategy for Row but now Column never has a dominant strategy. As we have seen, the $R_{12}$ swaps in this pipe always change the equilibrium conditions. Conversely, the $C_{12}$ swaps have no effect. Of thc 16 $X_{34}$ swaps, 14 cause changes in equilibrium.

Table 6.2 summarizes the effect of swaps on equilibria for the four pipes. The other two pipes, labelled Anti-Alibi and Anti-PD, have no equilibrium changes caused by $X_{12}$ swaps. This means that the four games on each layer have the same equilibrium outcome. Since the outcomes are different on each layer, all the $X_{34}$ swaps must change the equilibrium outcome. The Anti-PD pipe is contained within the region where both players have dominant strategies and the four outcomes are the best combinations possible: (4,4), (3,4), (4,3) and (3,3). In Anti-Alibi, only Row has a dominant strategy, but the outcomes are equally efficient.

Figures 6.8 to 6.11 provide a reference set of order graphs for the games in each of the the four named pipes.

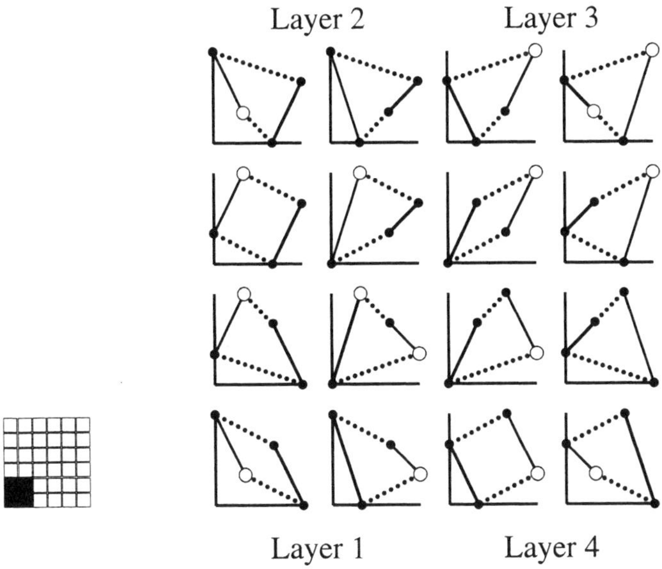


Figure 6.8: The PD-pipe (The Microcosm)

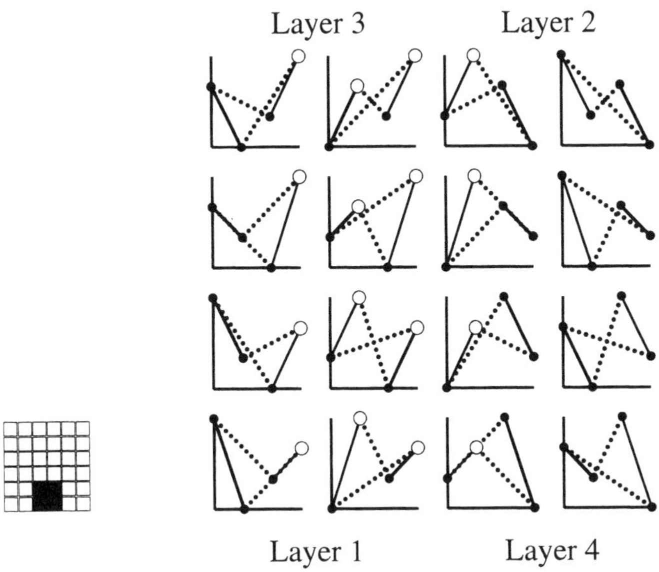


Figure 6.9: The APDF-pipe

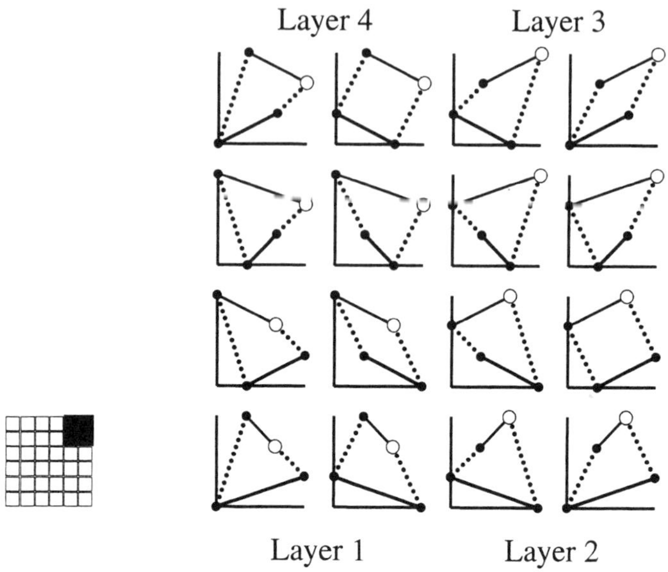


Figure 6.10: The Anti-PD pipe

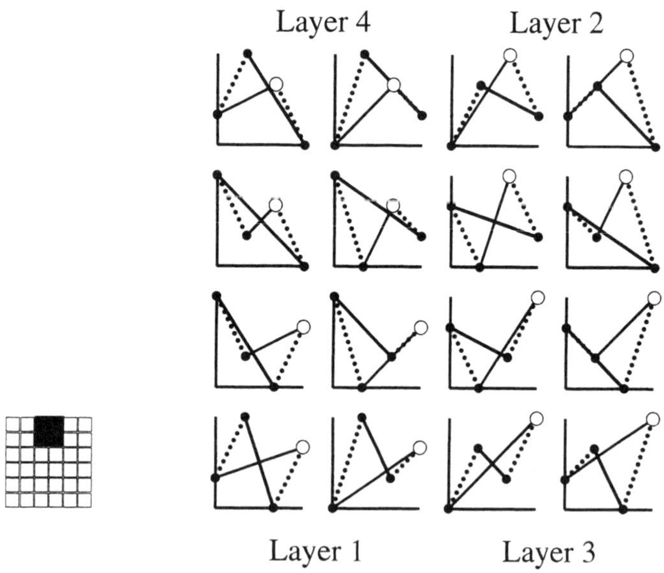


Figure 6.11: The Anti-APDF pipe

# Chapter 7

# 37 Holes, Coordination games, Battles of the Sexes and the hotspots

The strangest feature in the graph of the $2 \times 2$ games ties together a remarkable collection of issues. The rambling title of the chapter just begins to suggest the list of topics related to hotspots.

Hotspots are eight-game subspaces with the same generator set as the pipes we looked at in the last chapter. The fact that one set of generators produces groups with 16 members and groups with eight implies some games are fundamentally different from others.

It is as if we mixed Roman and Arabic numerals in different proportions on a single clock and defined "plus one" to mean "jump to the next member of the same set of numerals". Starting with a Roman numeral always yields a Roman numeral; starting with an Arabic numeral always yields a cycle of Arabic numerals. The groups are defined by the same generator set but have different numbers of members.

The mechanism that distinguishes hotspots and pipes is different, however. The operator does not distinguish among different types of games. Instead, patterns of payoffs in some games make combined operations equivalent, reducing the number of distinct games in a cycle. (See the subsection on rotations and reflections on page 23.)

In this chapter we present the structure of the hotspots, explain

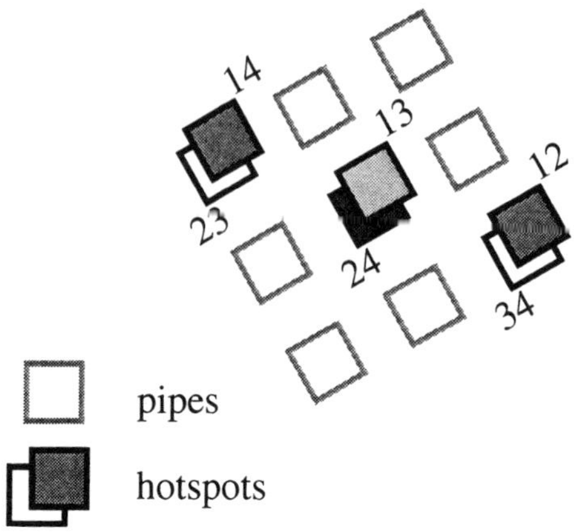


Figure 7.1: Hotspots and pipes

why they occur, and complete the topology. We then examine the games in several hotspots, including the Battle of the Sexes, Coordination games and "cycle" games.

## 7.1 Location and structure of hotspots

Figure 7.1 highlights the hotspots in the standard layout. Because each hotspot connects a distinct pair of layers, we can use layer numbers to identify each hotspot. The grey square marked "14" stands for a hotspot with one tile on Layer 1 and one on Layer 4. There are exactly six ways to combine two layers from a set of four and for each there is a hotspot.

Games in hotspots share the property that applying $C_{12}R_{12}$ or $C_{34}R_{34}$ (i.e., symmetric swaps) produces the same game. Figure 7.2 illustrates the two sequences for the transformation of $g_{261}$ to $g_{252}$ in hotspot 23. The order graph of $g_{261}$ has undergone different manipulation along each route but the resulting order graphs are identical. The corresponding payoff matrices show that the assumption of equivalence under row and column exchange must hold. Figure 7.3 shows the linkages within a hotspot the same way that Figure 6.4 on page 98 presented the structure of the pipes.

Hotspots that share the same stacks are not close to each other

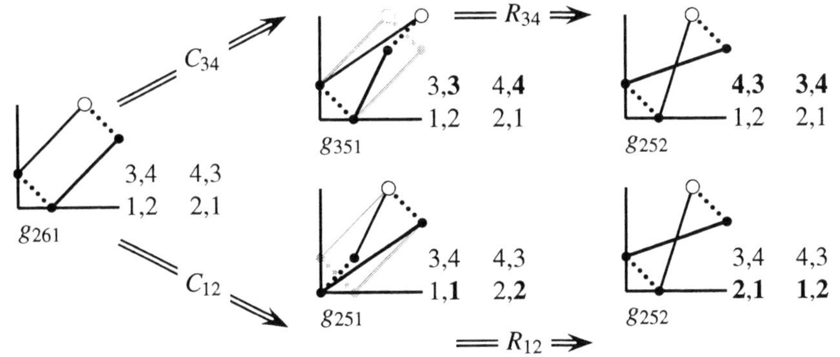


Figure 7.2: Equivalence of $C_{12}R_{12}$ and $C_{34}R_{34}$ transformations

in the graph. In a pipe, a game can be transformed to any other in the same stack by at most four swaps. It may take six swaps to reach a game in the complementary hotspot. Since the maximum path length between any two games in the graph is six, some games in the hotspot stacks are as far from each other as they can possibly be.

There are only four distinct types of hotspot. The central hotspots, 13 and 24, are distinct. Those labelled 14 and 12 are R&G reflections, as are 23 and 34.

As with a pipe, the graph of a hotspot can be rearranged so that no links cross. The graph that results will tile the plane in a "brick

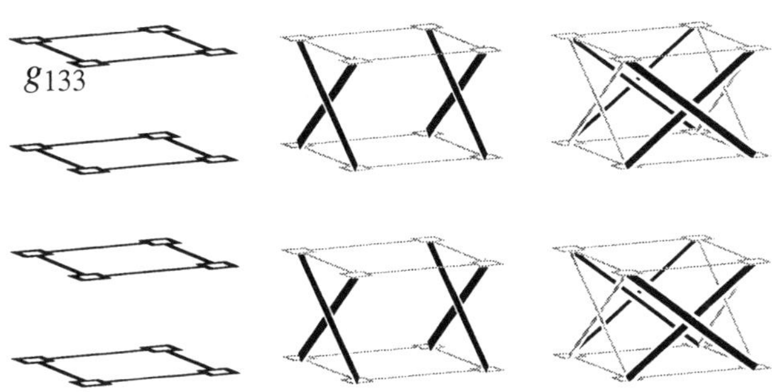


(a) (b) (c)

Figure 7.3: Linkages for two hotspots sharing a pile of tiles

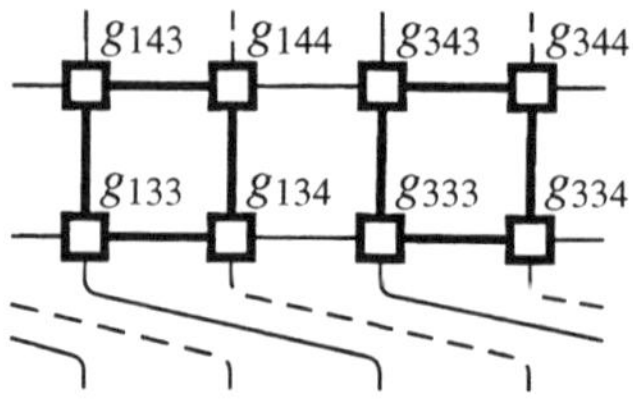


Figure 7.4: Hotspot 13 with the links untangled

wall" pattern. In Figure 7.4, hotspot 13 is shown with the tiles emphasized by thick lines. Game $g_{133}$ connects to $g_{334}$ and $g_{143}$ to $g_{344}$ by $C_{34}$ swaps horizontally. Game $g_{133}$ connects to $g_{343}$ and $g_{333}$ to $g_{143}$ by $R_{34}$ swaps. The $R_{34}$ links are offset two positions as they loop from bottom to top. The planar representation of the graph confirms that, like the pipe, the hotspot can be embedded in a torus.

## 7.2 How many holes? Thirty-seven

In the last chapter, on page 101, the six pipe-toruses were each attached to the four layer-toruses to produce a 25-holed surface. The six hotspot toruses can now be attached to this structure to finish the description.

Since a hotspot touches two layers, incorporating one into the surface will add two holes. One comes with the hotspot torus itself, and the other appears because the hotspot must form an arch connecting tiles on two layers (see Figure 7.5). The argument is essentially the same as the one for the pipes illustrated by Figure 6.6.

Since there are six hotspots, each adding two holes, the graph of the space of the $2 \times 2$ games can be embedded in a surface with $25 + 12 = \mathbf{37}$ holes. All the swaps are accounted for and the characterization of the topology is complete.

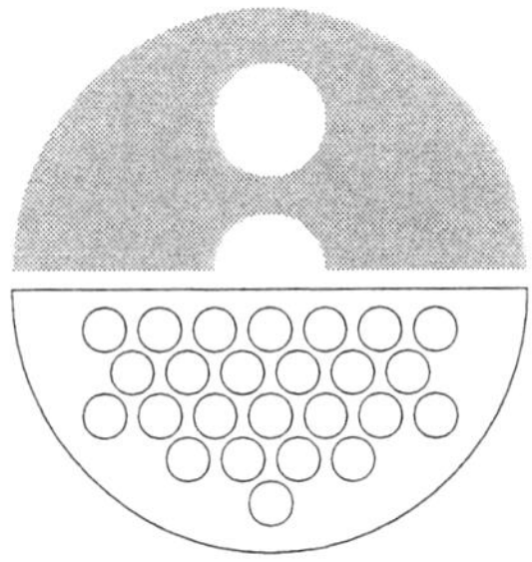

Figure 7.5: Adding a hotspot to 4 layers and 6 pipes

## 7.3 Hotspots and their games

Two hotspots are of special interest. The first, labelled 13, includes four Coordination games and four versions of the Battle of the Sexes. All the games have two Nash equilibria. The hotspot which shares the same stack, labelled 24, contains games with no Nash equilibrium. The games in these subspaces are "far" from each other but their order graphs bear a geometric resemblance: games with no equilibrium look like 90° rotations of games with two equilibria. Order graphs for the games in hotspots 13 and 24 are collected in Figure 7.6.

### 7.3.1 The two-equilibrium hotspot

The four games on the left of Figure 7.6(a) are variants of the Battle of the Sexes. The four on the right are variants of the Coordination game.

#### Battles of the Sexes

Battles of the Sexes have two Nash equilibria which are rank-sum equivalent but distributionally different. Rank-payoffs are either (3, 4) or (4, 3). Examples of the BoS in the literature generally set the

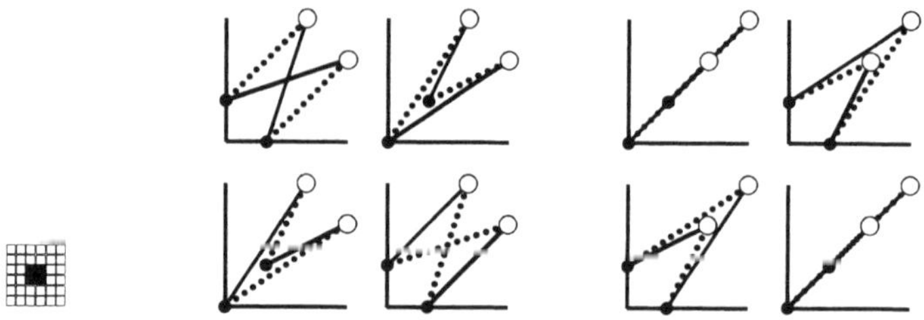

Layer 1 Layer 3

(a) Hotspot 13: Battle of the Sexes, Coordination

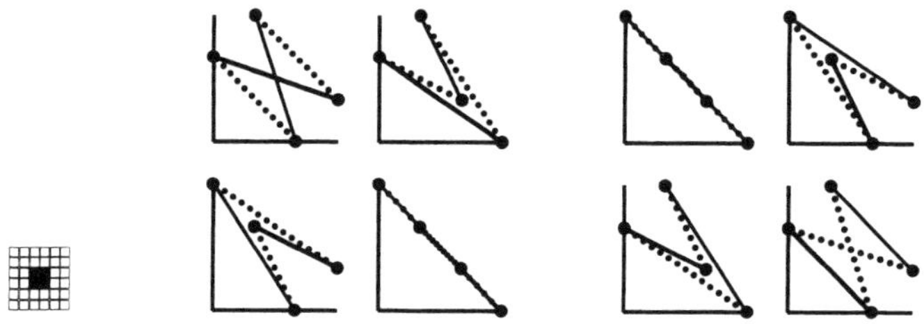

Layer 2 Layer 4

(b) Hotspot 24: games with no equilibrium

Figure 7.6: The two-equilibria and no-equilibrium hotspots

value of the two lowest-ranked payoffs to a common value. The effect is to produce an order graph like the central panel in Figure 7.7.

### Coordination games

Strict-ordinal symmetric Coordination games differ from the Battle of the Sexes in having multiple equilibria that are distributionally equivalent but differ in rank-sum. The difference in rank-sum is not generally the focus of attention. Indeed most writers appear to define the coordination problem as one in which, as Ullmann-Margolit [38] puts it, "there are several outcomes most preferred by all concerned." According to this definition there are no strict ordinal $2 \times 2$ Coordination games. Rapoport and Guyer [23] classify all of the games we call Coordination games as *no-conflict* games on the grounds that selecting a strategy can never present a serious problem for players.

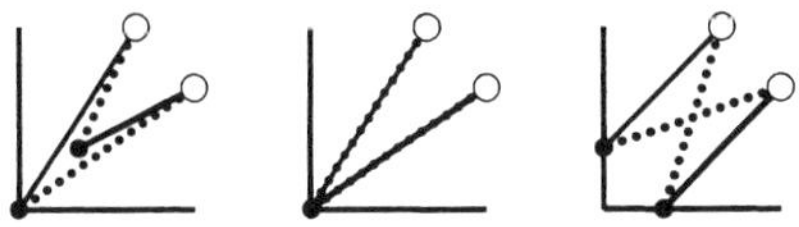

Figure 7.7: Variants of the symmetric and quasi-symmetric Battles of the Sexes

Requiring that coordination equilibria have precisely the same payoffs and that the outcomes be most preferred obscures much that is of interest in the coordination problem. In particular, it rules out the possibility of a low-level coordination equilibrium in which each player prefers to change strategy *providing all other players change strategy in a complementary way*. The games on the right of figure 7.6(a) all have this feature. When these games are included, the Coordination game can be used to model the problem of coordinating change. Abandoning the QWERTY keyboard for the more efficient Dvorak keyboard is a well known example modelled by the Coordination games in figure 7.6(a).

The most common representation of the Coordination game is a payoff matrix with 1s on the diagonal and zeros on the off-diagonal. It is the limit for all four of the games in the tile if we allow the two highest-ranked payoffs to go to 1 and the two lowest to go to zero for both players.

Schelling ([33] p. 83-9) proposed abandoning the then-traditional division into zero-sum and non-zero-sum games in favour of a categorization that emphasizes the continuum between pure conflict and pure common interest. The games in this group suggest that there is no unidimensional continuum from pure conflict to pure common interest. In both Coordination and Battle of the Sexes games, any move that improves the payoff for one player also improves the payoff for the other. The games are therefore, at least at the level of individual choices, games of pure common interest. There remains an element of conflict at the level of equilibrium choice, but there is none at the level of individual decisions at any position.

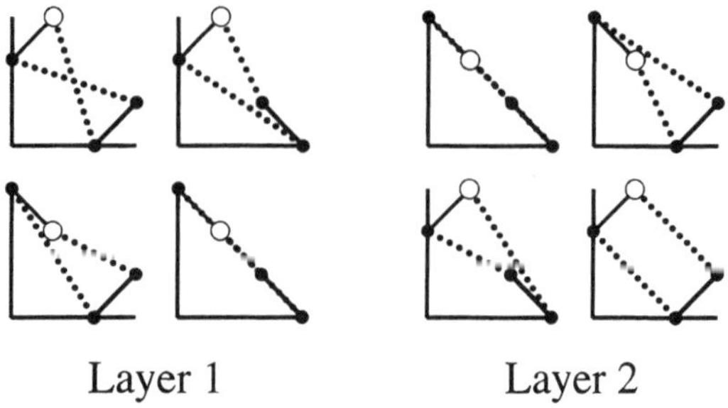


(a) Rewired versions of games with no Nash equilibrium

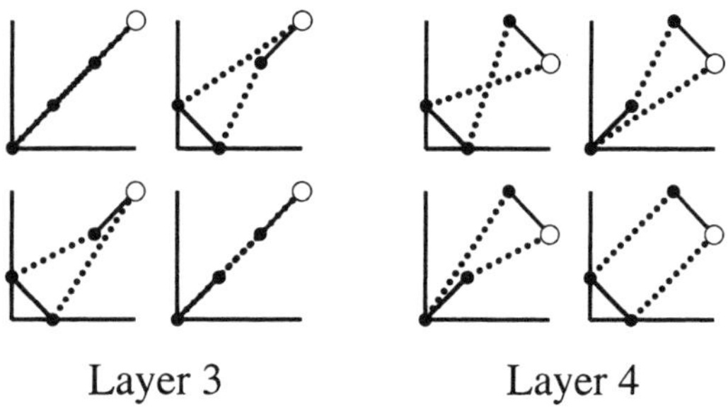


(b) Rewired versions of Coordination games and BoSs

Figure 7.8: The other hotspots: (a)12, (b)34

### 7.3.2 The no-equilibrium hotspot

The other hotspot at the centre of the standard layout consists of games which are essentially 90° rotations of the Coordination and Battle of the Sexes games. Since those games were all games of common interest in the sense that the inducement correspondences were positively sloped, these are all games of total conflict, having inducement correspondences that are all negatively sloped.

The graphical feature that ensures that none of these games have equilibria is that at every position, negatively sloped inducement correspondences lead away in the same direction, ensuring that if one player likes the position, the other has an improving move. The sequence of best response choices cycles around the order graph, and the games are sometimes called *cycle games* [23].

### 7.3.3 The other hotspots

The remaining hotspots are interesting primarily in the way they differ from the two already examined. Two hotspots, 12 and 34, are illustrated in Figure 7.8. The R&G reflections, 14 and 23, are not shown because they add no information.

Each game is a rewiring of a game in the previous groups: the payoff pairs are unchanged but their arrangement in the payoff matrix differs. Half of the games are quasi-symmetric. This feature is examined in the following chapter.

The games have unique Nash equilibria, none of which are especially interesting. The hotspots lie on the instability edges but outside the instability zone discussed in Chapter 6, where $X_{12}$ swaps change best responses. In every game, Column has a dominant strategy, and in the games on the lower half of each tile, both players have dominant strategies. The most significant feature may be the number of games with very unfair equilibria.

## 7.4 Geography of the social dilemmas

Just as there are asymmetric Prisoner's Dilemmas, there are also asymmetric Coordination games shown in Figure 7.9 that lie adjacent to the symmetric Coordination games. With Stag and Hare, which lies on the symmetric diagonal, they form a border on two sides of the Coordination tile.

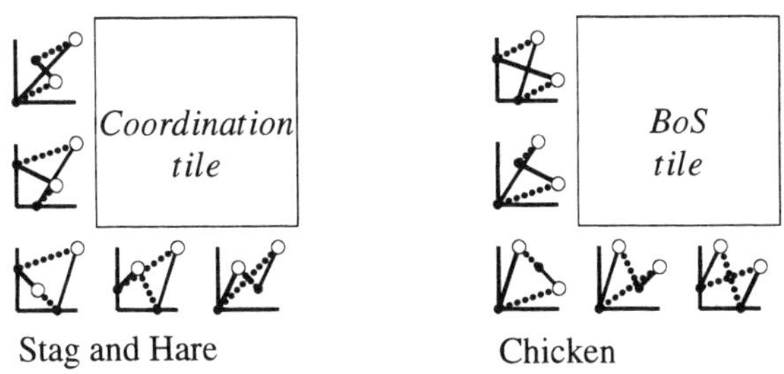


Figure 7.9: Chicken and asymmetric BoS; Stag and Hare and asymmetric Coordination games

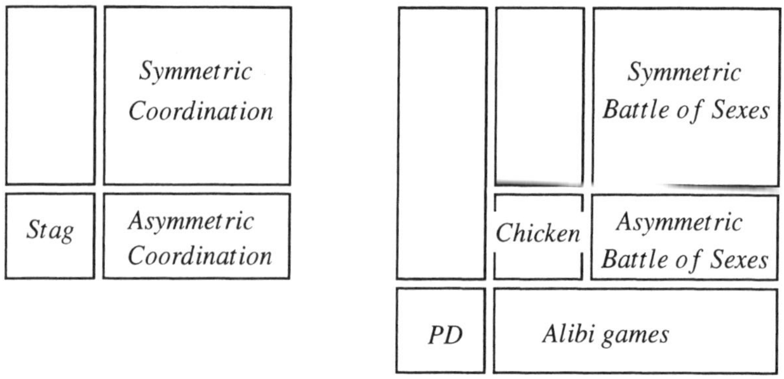


Figure 7.10: Asymmetric, symmetric (and quasi-symmetric) dilemmas

The Battles of the Sexes tile has a similar asymmetric border. It consists of asymmetric BoS games with Chicken at the corner on the diagonal. Chicken and Stag and Hare, it appears, are the extreme cases of Battle of the Sexes and Coordination games respectively. The resulting $3 \times 3$ region is in turn half-bordered by the Prisoner's Dilemma Family as Figure 7.10 shows.

This remarkable region is exactly the one introduced in Chapter 5 and visualized in Figure 5.5 on page 80. These are the social dilemmas. We established in Chapter 5 that these games were a connected subset. Now we have categorized them and shown how they are related, using the concept hotspots. Figure 7.10 is an abstract version of Figure 5.5 (p. 80). It is a useful roadmap to the territory of the social dilemmas.

# Chapter 8

# Classifying conflict

## 8.1 Conflict, no conflict, mixed interests

Game theory is regularly defined and marketed as the study of conflict, with titles such as *Game Theory: Analysis of Conflict*, by Myerson [19], *Game Theory: Mathematical Models of Conflict*, by Jones [14], *Game Theory as a Theory of Conflict Resolution*, by Rapoport [22], *The Strategy of Conflict*, by Schelling [33], *Conflict Among Nations: Bargaining, Decision Making and System Structure in International Crises*, by Snyder and Diesing [35] and *The Structure of Conflict*, edited by Paul Swingle [37].

The emphasis on conflict is reflected in definitions of the field:

> Game theory is a branch of mathematical analysis developed to study decision making in conflict situations.
>
> Dr. Francis Heylighen [13]

> Game theory is the interdisciplinary study of conflict.
>
> Dr. Daniel King [15]

> Game theory studies formal models of conflict and cooperation.
>
> Dr. Bernhard von Strengel [40]

Game theory did in fact originate in the analysis of games of pure conflict: *Theory of Games and Economic Behavior* is based on a solution to the zero sum games[1]. By 1958, however, Schelling was calling for more attention to non-constant sum games. "Pure conflict, in which the interests of two antagonists are completely opposed, is a special case" ([33] p. 4).

> On the strategy of pure conflict – the zero-sum games – *game theory* has yielded important insight and advice. But on the strategy of action where conflict is mixed with mutual dependence – the nonzero-sum games involved in wars and threats of war, strikes, negotiations, criminal deterrence, class war, race war, price war, blackmail, maneuvering in a bureaucracy or in a traffic jam, and the coercion of one's own children – traditional game theory has not yielded comparable insight or advice ([33] p. 83).

Schelling set out to "enlarge the scope of game theory, taking the zero-sum game to be the limiting case rather than a point of departure". The other limiting case, in his view, was the Coordination game.

> The essentials of a classification scheme for a two-person game could be represented on a two-dimensional diagram... All possible outcomes of a pure-conflict game would be represented by some or all of the points on a negatively inclined line, those of a pure common-interest game by some or all of the points on a positively inclined line. In the mixed game, or bargaining situation, at least one pair of points would denote a negative slope and at least one pair a positive slope ([33] p. 88).

[1] Von Neumann and Morgenstern write "While these games are not typical for major economic processes, they contain some universally important traits of all games and the results derived from them are the basis of the general theory of games" ([39] p. 34).

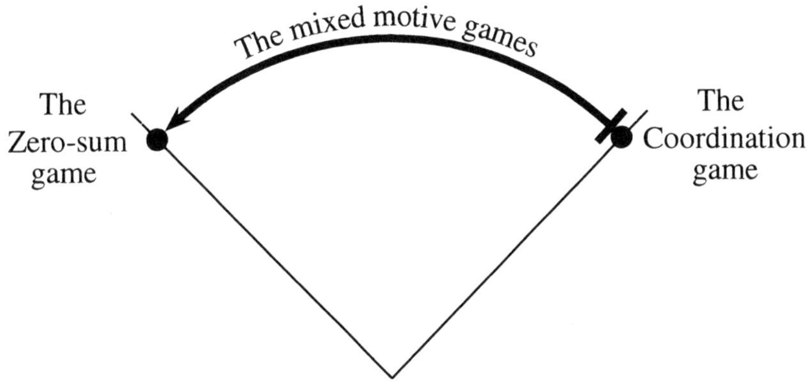


Figure 8.1: A graphical view of Schelling's classification scheme

A version of the model Schelling describes in the quotation is shown as Figure 8.1. A more complete model incorporating some of the results in this chapter appears later as Figure 8.9.

Schelling himself provided one of the most satisfying discussions of the mixed motive games. By 1970 Anatol Rapoport would write, "It seems to me that the real value of game theory... lies in the subsequent development of the theory beyond the context of the two-person constant sum game"[22](p. 38).

Classification according to the degree of conflict quickly became standard procedure. In 1966 Rapoport and Melvin Guyer published a taxonomy of the $2 \times 2$ games in which the largest category was *no-conflict games*. In 1976 Rapoport, Guyer and David Gordon [24] revised the classification, making *No-conflict Games*, *Mixed-motive Games* and *Games of Complete Opposition* the primary categories.

Even so, relatively few mixed motive games have been examined. In *The Strategy of Conflict*, Schelling explicitly analysed only seven. Although he suggested a classification scheme, he did not provide a systematic classification and no satisfactory classification of the so-called mixed motive games has emerged.

This chapter presents a systematic classification based on the topology of the $2 \times 2$ games. It also provides an improved terminology for discussing conflict in the $2 \times 2$ games, new subdivision of the games, and a map of the relationships among games based on the degree of conflict.

## 8.2 Describing conflict using inducement correspondences

Games of pure conflict, pure common-interest and mixed motives can be described in terms of the slopes of the inducement correspondences. In a game of pure conflict every inducement correspondence is negatively sloped. Any action that improves the outcome for one player must make the outcome worse for the other. In a game of pure common-interest, every inducement correspondence is positively sloped.

Mixed-motive games, to use Schelling's term, can have one, two or three positively sloped inducement correspondences. There are therefore five levels of conflict for $2 \times 2$ strict ordinal games. The situation with two positively sloped inducement correspondences can be further divided into the case in which both players have mixed slopes and the case when one player has positive slopes and one has negative slopes. This latter situation is significantly different from the other mixed motives games, as we will show.

Using inducement correspondences seems natural, but it results in a classification that differs from Schelling's and others'. If a hard distinction is made between zero-sum games and nonzero-sum games, as Schelling does, the Prisoner's Dilemma and the six zero-sum games are qualitatively distinct although they may be quantitatively similar. If the line is drawn between games in which all inducement correspondences are negatively sloped and those with one positive slope, as we suggest, the constant-rank-sum games are grouped with the Prisoner's Dilemma. Arguably the PD is more like the constant-rank-sum game $g_{161}$, which has a single Nash equilibrium, than $g_{161}$ is like the constant-rank-sum game $g_{234}$, which has no Nash equilibrium. The point is that grouping games on the basis of the slope of the inducement correspondences provides a coherent but distinct approach.

The no-conflict games are also treated differently using the inducement correspondence approach. Schelling asked "The zero-sum game is the limiting case of pure conflict, what is the other extreme?" and answered, "It must be the 'pure-collaboration' games in which the players win or lose together... they must rank all outcomes iden-

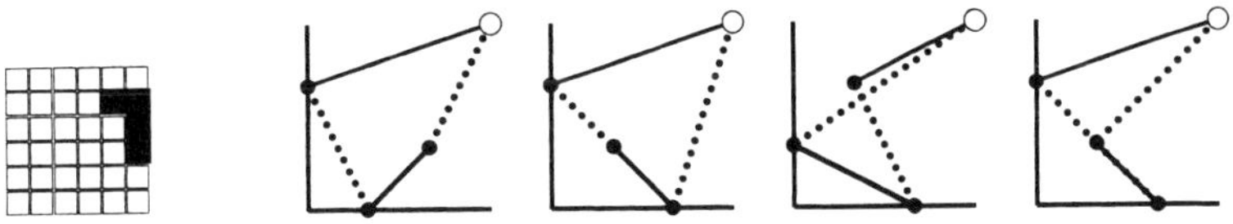


Figure 8.2: "No-conflict" games that are also mixed motive games

tically" ([33] p. 84). There are six ordinal games that fit this description. Four have a single Pareto-optimal equilibrium and two have two Nash equlibria, suggesting that the category is not internally consistent.

There are 10 other games with positively sloped inducement correspondences, some with one and some with two equilibria. Are the games with two equilibria and positively sloped inducement correspondences most like other games with two equilibria or most like games in which the players have common rankings? Topologically they are closer to the related Coordination games than to other common-ranking games, suggesting that Schelling's classification is externally inconsistent as well.

There is another terminological issue that can muddy the water. Rapoport and Guyer [23], Rapoport, Guyer and Gordon [24], and Brams [5] all classify payoff matrices in which one cell contains a (4,4) payoff as "no-conflict" games. Players can achieve their best payoffs simultaneously. These games form Layer 3. By their definition, the 36 "no-conflict" games include 10 games with two negatively-sloped inducement correspondences and 16 with one. In other words, most of the no-conflict games are mixed-motive games under Schelling's widely accepted classification. Figure 8.2 provides four examples.

## 8.3 A single-surface map of the 144 games

The patterns of conflict and common interest spread across layers. A new, single-layered, figure is helpful for examining the conflict patterns.

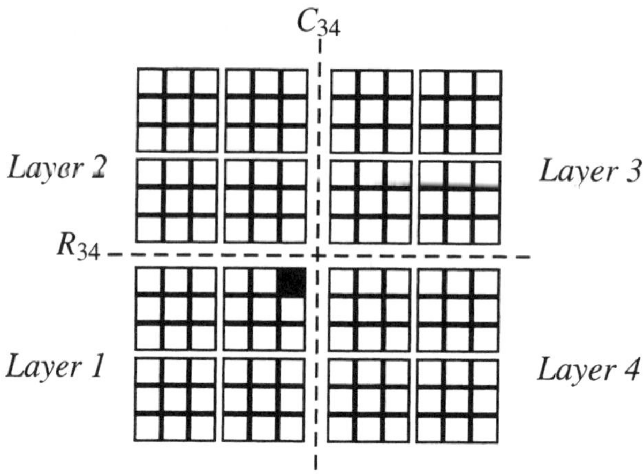


Figure 8.3: Connecting four layers at the instability boundaries

Links between layers occur where there are both $X_{12}$ and $X_{34}$ swaps, so layers can only be connected along the boundaries between Rows (or Columns) 1 and 2, 3 and 4, or 5 and 6. If we lay out each layer so the edges follow these boundaries, the layers can be joined edge to edge.

There are nine ways to paste the layers together based on which boundaries form the row and column edges. Three are symmetric, and one of these maintains the relative positions of the four layers that we have used so far. In Chapter 6, Section 6.2 (page 95) we described the two *instability zones* that run through the Prisoner's Dilemma pipe and mark the boundary between regions with and without dominant strategies. If the torus for each layer is cut along the instability boundaries (between Rows 1 and 2 and Columns 1 and 2), games from the Prisoner's Dilemma stack are in the top right corner. Along these edges the layers can be joined by $X_{34}$ swaps into a single toroidal surface. The resulting figure emphasizes the centrality of the Prisoner's Dilemma and captures the relationships between layers that we uncovered in Chapter 5 on the Prisoner's Dilemma and the Alibi games.

The resulting $12 \times 12$ planar graph (Figure 8.3) includes all 144 games. It conserves the links within layers and shows 24 more con-

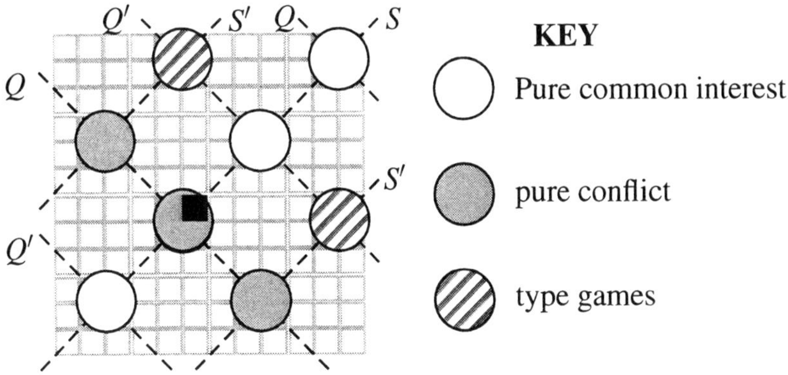


Figure 8.4: The pure groups and the symmetry axes

nections than the four-layered version[2]. In total, 336 of the 432 edges of the graph of the $2 \times 2$ games are directly accessible from the diagram. Furthermore it can be rolled into a single torus that includes all 144 games.

The new figure allows us to visualize patterns that cross layers. The symmetric games form the positive diagonal, $S$, and the quasi-symmetric games form a pair of negative diagonals, $Q$ and $Q'$. On Layers 2 and 4, there is an additional diagonal, $S'$, consisting of games for which individual payoff patterns are the same patterns as in the symmetric games but are oriented differently to each other. The diagonals in Figure 8.4 are intimately connected with the distribution of common- and opposed-interest games.

## 8.4 The pure cases

The principle diagonals can best be understood as *Villarceau circles*. Exactly four topologically distinct circles can be drawn through any point on a torus[43]. One is in the plane of the torus and passes around the hole like a bead of icing on the top of a doughnut. The second is perpendicular to the first and can be pictured as a piece of

[2]In Chapter 9 this configuration is called the Periodic Table of the $2 \times 2$ games.

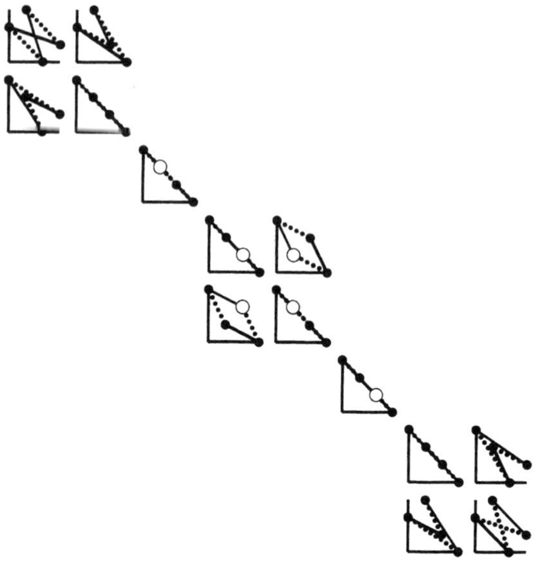

Figure 8.5: The pure conflict games

string tied through the hole. Two more, the Villarceau circles [3], run on the diagonal, winding around the doughnut and through the hole, one clockwise and one counterclockwise. Villarceau circles through a given point intersect twice.

The axes of symmetry and quasi-symmetry form two pairs of Villarceau circles that intersect at the eight places indicated in Figure 8.4. The clusters of four games at the intersections are pure cases of common interest or conflict. There are three types

1. Grey circles mark three clusters of games with four negatively sloped inducement correspondences. These are the **pure conflict games**. The three clusters are joined by a band of constant-rank-sum games lying along a quasi-symmetric axis. Figure 8.5 shows order graphs for the 14 pure conflict games. The group includes the Prisoners's Dilemma, eight games with no pure-strategy Nash equilibrium and six constant rank-sum games.

[3] For a demonstration that they really are circles, see Weisstein [43].

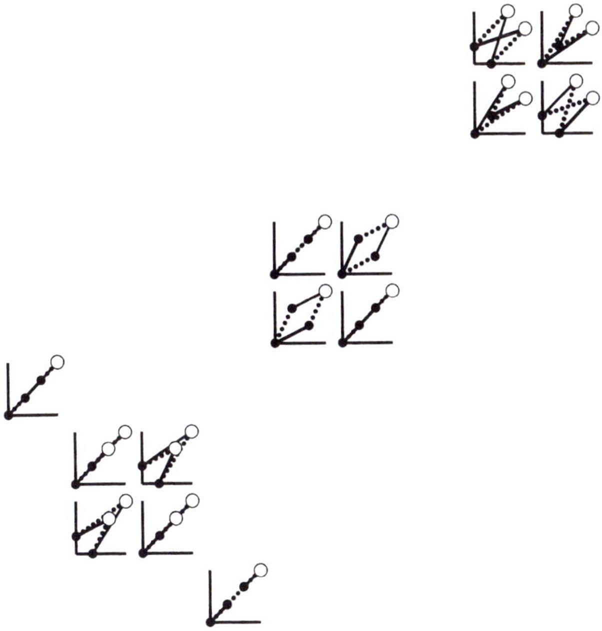

Figure 8.6: The pure common-interest games

2. White circles in Figure 8.4 mark three clusters of games with four positively sloped inducement correspondences. These are the **pure common-interest** games. They fall along the symmetric diagonal. There are two pure common-interest games adjacent on a quasi-symmetric axis to one of the clusters. Figure 8.6 shows the order graphs for the the 14 games in this group. The group includes Coordination games, Battle of the Sexes games and six common ranking games (or Schelling's "pure-collaboration" game([33]p. 84)).

3. Grey and white striped circles in Figure 8.4 indicate games which are **pure conflict** for one player and **pure common interest** for the other. This is a new class of game that we call the "Type" games because the players are of different unmixed types. Figure 8.7 shows order graphs for one of the two clusters of Type games. (The other cluster is an R&G reflection of the one shown.)

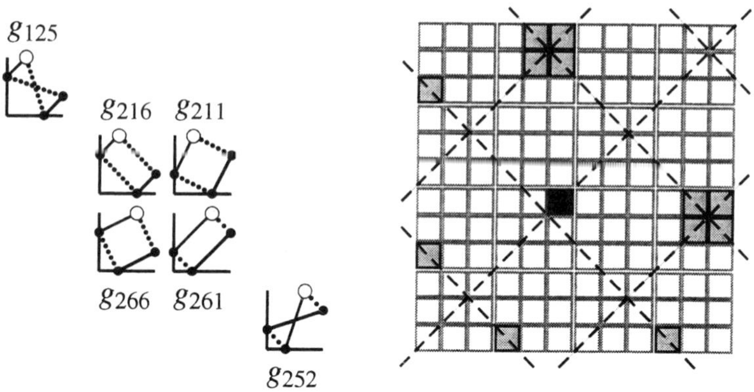


Figure 8.7: Pure Type games

## 8.5 Giver and taker: the Type games

In Type games, the players are asymmetric in the strongest way. One player always gains by making the other better off; the other always loses by making her partner better off. Unlike Chicken and Stag Hunt, in which both players have mixed interests, the motives of players in these Type games are absolutely unmixed.

In Type games the players live in different moral universes. One type is never led into temptation, the other is never free of temptation. One needs no moral instruction, the other must be restrained by law. One freely casts his bread upon the water and the Lord provides, while the other must live by theft. These games seem perfectly suited for exploring a whole class of morally ambiguous situations – cases in which agents may debate morality from fundamentally different material situations.

Type games have appeared in the literature from time to time. Schelling [33], for example, used $g_{125}$ to describe the situation of a blackmailer and his victim[4]. Licht [16] uses two examples to describe situations in which two nations with very different powers

[4]Interestingly he imagines the blackmailer unilaterally agreeing to compensate his victim in a certain situation. The result is equivalent to a $C_{23}$ swap which changes the game to $g_{124}$ ([33] p. 159).

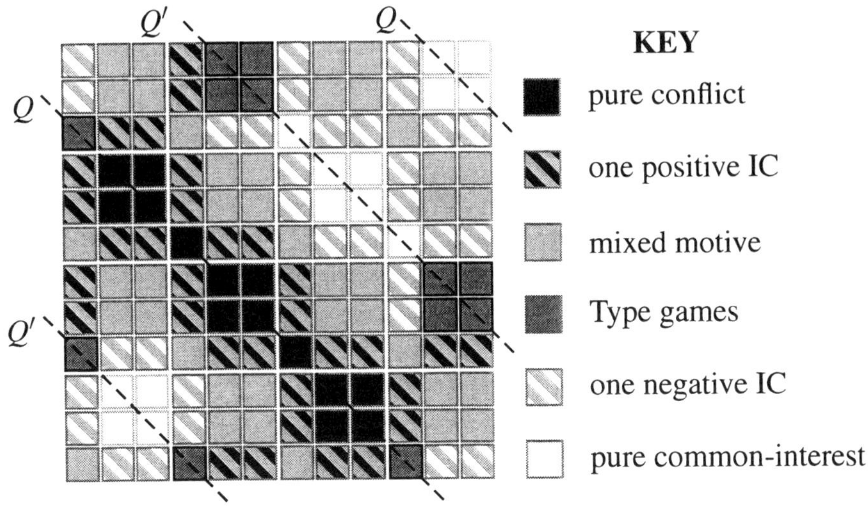


Figure 8.8: Conflict and common interest

negotiate. His *Ideological Hegemony 1* and *2* are reassigned versions of $g_{211}$ and $g_{261}$.

## 8.6 Mixed motive games

Figure 8.8 shows all games shaded according to whether they have $0, 1, 2, 3$, or 4 negatively sloped inducement correspondences. Pure conflict games are black and pure common-interest games are white. The rest of the games are mixed motive games in Schelling's sense [33].

The mixed motive region includes, among others, Chicken, Stag Hunt, four games with no equilibrium and the Alibi games described in Robinson and Goforth[26]. Hatched bands around the pure conflict and pure common-interest games show where games have one positively or one negatively sloped inducement correspondence. The dark and light gray regions consist of games with two positively sloped inducement correspondences and two negatively sloped. In the light gray games, each player has one positively sloped and one negatively sloped inducement correspondence. The dark gray games are the Type games.

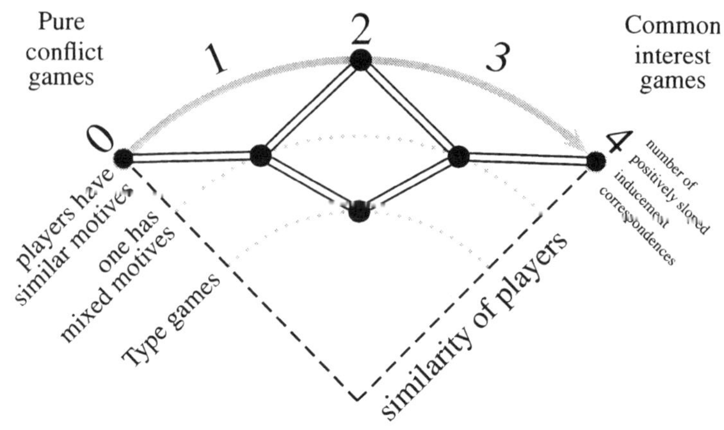


Figure 8.9: A discretely corrected version of Schelling's classification scheme

## 8.7 Extending Schelling

The number of positively and negatively sloped inducement correspondences is a useful measure of the degree of conflict or common-interest in the ordinal $2 \times 2$ games. It also provides a complete classification that has significant advantages over the model proposed by Schelling in 1958 [34]. A classification by inducement correspondence allows us to distinguish between mixed motive games and the set we call Type games.

Figure 8.9 shows how Schelling's proposal can be extended to include different types of mixed-motive ordinal games. In Figure 8.9 there are five degrees of alignment of interests, and three degrees of mixed motives. Along the upper arc there are three cases in which the players are completely similar in the likelihood that they will face negatively or positively sloped inducement correspondences. Below the arc are cases in which players' incentives differ. The type games on the lowest path exhibit complete differentiation.

## 8.8 Completing the classification

An approach that is more consistent with the topological structure distinguishes row and column players. So far we have treated the

Figure 8.10: Number of negatively sloped inducement correspondences for Row and Column

case in which Row has two negatively sloped inducement correspondences and Column one as equivalent to a case in which Column has two and Row has one. We have therefore implicitly assumed that players are indistinguishable, which is the assumption that Rapoport and Guyer [23] used to reduce the number of games from 144 to 78. If players are not interchangeable there are three degrees of common interest for each player, making a total of nine possible cases.

In Figure 8.10, pure common-interest games for Row (both her inducement correspondences are positively sloped) appear as white squares in the lower left figure. Pure conflict games for Row appear as black squares and games in which Row has mixed motives are grey[5]. Column's motives are shown with the same colours on the upper left figure. Only half of each square is coloured for Column.

When the pattern for Column is laid over the pattern for Row

[5] The nine-game blocks are offset by one column from the blocks of the dominant strategy layout of Figure 5.2on page 77. There is a one-row displacement for Column. The pure conflict, pure common-interest, Type and pure mixed-interest clusters occur where the blocks "match".

Figure 8.11: How conflict and common-interest games tile the plane

the result is the remarkable pattern on the right of Figure 8.10. Pure conflict games are all black, and pure common-interest games are entirely white. In the grey games both players have mixed motives. Black and white games are games in which players have different kinds of motivation: the Type games.

All the games with black or white – where individual players have unmixed motives – cluster around the intersections of the Villarceau circles marking the symmetric and quasi-symmetric games.

The overall pattern is even more striking if multiple copies of Figure 8.10 are laid edge to edge, producing Figure 8.11. Tiling the plane in this way brings out features that wind around the torus. Figure 8.11 can read like a topological map. In the toroidal world of the $2 \times 2$ games there are black highlands where conflict prevails and white lowlands where cooperation is the rule.

## 8.9 Structure of conflict

In this chapter we have carried forward a project begun by Schelling in 1958. We propose a complete classification of the ordinal $2 \times 2$ games in terms of the degree of conflict among the players. We have shown that the 1966 taxonomy developed by Anatol Rapoport and Melvin Guyer provides an inconsistent and misleading treatment of conflict in the $2 \times 2$ games.

Topological relationships and symmetry allow us to uncover unexpected relationships among games of pure conflict and games of pure common interest. The fact that the pure conflict and common-interest games exhibit such a clear pattern demonstrates the organizing power of the topological approach. Our treatment of the mixed motive games reveals a class of games, the Type games, that has not previously been recognized as distinct. In Type games the players live in different moral universes. These are games that will surely interest moral philosophers.

# Chapter 9

# A Periodic Table for the 2×2 Games

The Periodic Table was one of the great discoveries in the field of chemistry. It includes every possible element, ordered by atomic weight and chemical properties. The first successful versions were developed in the 1860s, before all the elements had been observed and before chemists understood the underlying structure of the atom. Identifying periodic properties allowed chemists to predict the properties of elements that had not yet been discovered. The table rapidly became an essential tool for research and for teaching.

The relationships among the elementary $2 \times 2$ games are more complex than the relationships among the chemical elements. The periodic Table of Elements begins with a simple ordering by weight. In addition to being a structural feature, weight provides a natural numbering system for the elements. The Periodic Table then identifies a cycle with an increasing period running from base to acid. Each cycle becomes a row of the table.

For $2 \times 2$ games, both the number of games and the underlying structure are completely known and the difficulty lies in the complexity of the structure. Where each chemical element has at most four neighbours, each $2 \times 2$ game has six neighbours. Where the elements are linearly ordered by atomic weight, each game is embedded in several cyclical relationships. Where atomic weight provides a natural numbering scheme for the elements, there is no game that is obviously first. Where the graph of the Periodic Table of Elements has an Euler number of 2 and can be embedded in a plane, the graph of the $2 \times 2$ games has an Euler number of $-72$ and requires a 37-holed surface to represent all the links without crossing.

To project the $2 \times 2$ games onto a plane many links have to be left out. The problem is to find the most informative projection.

In this chapter we propose a Periodic Table for the $2 \times 2$ games. The table can be used to display many of the relationships described in previous chapters in a single two-dimensional chart, including symmetry relationships, dominance solvability, a systematic classification according to the degree of common interest or conflict, the relationships between conflict and dominant strategies, and between dominant strategies and the the presence of Nash equilibria, Pareto-optimality, the relationships among social dilemmas, and some topological features. In the process we illustrate and provide a critique of the typology introduced by Rapoport and Guyer [23] and developed further with Gordon.

There are in fact nine ways to produce a single torus by pasting layers together along $X_{34}$ boundaries. The version introduced on page 144 has several advantages; it places the Prisoner's Dilemma in a central position and it emphasizes the symmetric games by placing them on the diagonal of the figure. It preserves the relationships within layers developed in Chapter 3 and most of the symmetry relations described in Chapters 3 and 4. It displays the games with Pareto-inferior equilibria as the connected region identified in Chapter 5, and it incorporates the information about conflict and common interest developed in Chapter 8.

Because the arrangement highlights the economically and behaviourally significant games it is, in our view, the best candidate to serve as a Periodic Table of the $2 \times 2$ games.

## 9.1 The Periodic Table of the $2 \times 2$ games: indexing

Figure 9.1 shows the index numbers for the games in the Periodic Table. We have retained the topologically-based numbering introduced in Chapter 2.

| | 2 | 3 | 4 | 5 | 6 | 1 | 2 | 3 | 4 | 5 | 6 | 1 |
|---|---|---|---|---|---|---|---|---|---|---|---|---|
| 1 | 212 | 213 | 214 | 215 | 216 | 211 | 312 | 313 | 314 | 315 | 316 | 311 |
| 6 | 262 | 263 | 264 | 265 | 266 | 261 | 362 | 363 | 364 | 365 | 366 | 361 |
| 5 | 252 | 253 | 254 | 255 | 256 | 251 | 352 | 353 | 354 | 355 | 356 | 351 |
| 4 | 242 | 243 | 244 | 245 | 246 | 241 | 342 | 343 | 344 | 345 | 346 | 341 |
| 3 | 232 | 233 | 234 | 235 | 236 | 231 | 332 | 333 | 334 | 335 | 336 | 331 |
| 2 | 222 | 223 | 224 | 225 | 226 | 221 | 322 | 323 | 324 | 325 | 326 | 321 |
| 1 | 112 | 113 | 114 | 115 | 116 | 111 | 412 | 413 | 414 | 415 | 416 | 411 |
| 6 | 162 | 163 | 164 | 165 | 166 | 161 | 462 | 463 | 464 | 465 | 466 | 461 |
| 5 | 152 | 153 | 154 | 155 | 156 | 151 | 452 | 453 | 454 | 455 | 456 | 451 |
| 4 | 142 | 143 | 144 | 145 | 146 | 141 | 442 | 443 | 444 | 445 | 446 | 441 |
| 3 | 132 | 133 | 134 | 135 | 136 | 131 | 432 | 433 | 434 | 435 | 436 | 431 |
| 2 | 122 | 123 | 124 | 125 | 126 | 121 | 422 | 423 | 424 | 425 | 426 | 421 |

Figure 9.1: The Periodic Table of the $2 \times 2$ games

It is worth reviewing the information that the indices provide.

1. Games on a layer begin with the same first digit.

2. Games on a row share the same second digit, even when the sequence of row numbers is displaced. The same row sequence applies in each layer.

3. Games in a column share the same third digit, even when the sequence of column numbers is displaced. The same column sequence applies in each layer.

4. Games with the same row and column index are in the same stack.

5. Games with row index equal to column index on Layers 1 and 3 occupy the principle positive diagonal of the periodic table. These are the symmetric games, invariant under $R\&G = R^{\nwarrow}A$.
   - Folding the table along the positive diagonal will bring every game to coincide with its R&G reflection.
   - The reflection has the row and column index reversed.
   - Layers 2 and 4 fold onto each other: $g_{436}$ folds onto $g_{263}$.
   - Layers 1 and 3 each fold onto themselves: $g_{336}$ folds onto $g_{363}$.
6. Games with row index equal to column index on Layers 2 and 4 occupy the principle positive diagonal of Layers 2 and 4. These are the games that are invariant under $R^{\swarrow}A$.
7. Games with row index equal to $(7 - (column\ index))$ are quasi-symmetric. They occupy negative diagonals and are invariant under both $R^{\nwarrow}$ and $R^{\swarrow}$.
8. Each layer can be regarded as a torus. On Layer 2 for example, $g_{252}$ is adjacent to $g_{251}$ and $g_{224}$ is adjacent to $g_{214}$. $g_{222}$ is adjacent to both $g_{221}$ and $g_{212}$.
9. Layers are connected only between index 1 and 2 for row or column.
10. The full $12 \times 12$ layout can be regarded as a torus. $g_{252}$ is also adjacent to $g_{351}$. The edge connections are between index 1 and 2 for row or column.

In Figure 9.1 the shaded games are all members of the PD pipe described in Chapter 6. The 16 games in the pipe comprise four stacked tiles, one on each layer.

### On the choice of the numbering system

Selecting a numbering system is essentially an aesthetic problem. We have chosen to stay with the numbering for the standard layout because that layout provides the most natural approach to the

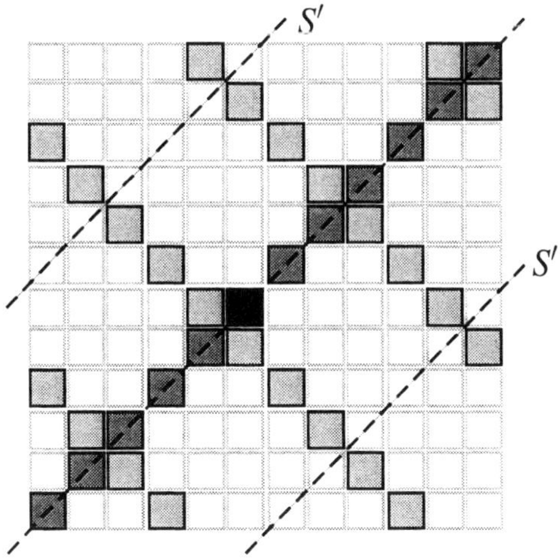


Figure 9.2: Symmetric and quasi-symmetric Axes

topological relationships. It distinguishes layers, which are a natural organizing device. It places symmetric and quasi-symmetric games on the principle positive and negative diagonals of the layers, which emphasizes the symmetries among games.

We begin numbering at the Prisoner's Dilemma for several reasons. All $2 \times 2$ games are equal in the the toroidal space occupied by the $2 \times 2$ games. The Prisoner's Dilemma, however, is the first among equals, being without question the most famous game of all. It also lies on the symmetric axis between the inter-layer boundaries and one of the quasi-symmetric axes[1]. As a result it is a natural reference point.

## 9.2 Axes of symmetry

The space of the $2 \times 2$ games exhibits several important symmetries. Figure 9.2 shows 11 symmetric games in dark grey and the Prisoner's Dilemma in black. The twelve games lie on the positive diagonal of the periodic table. These are the only games that do not

[1] Only the very undistinguished anti-PD shares this feature.

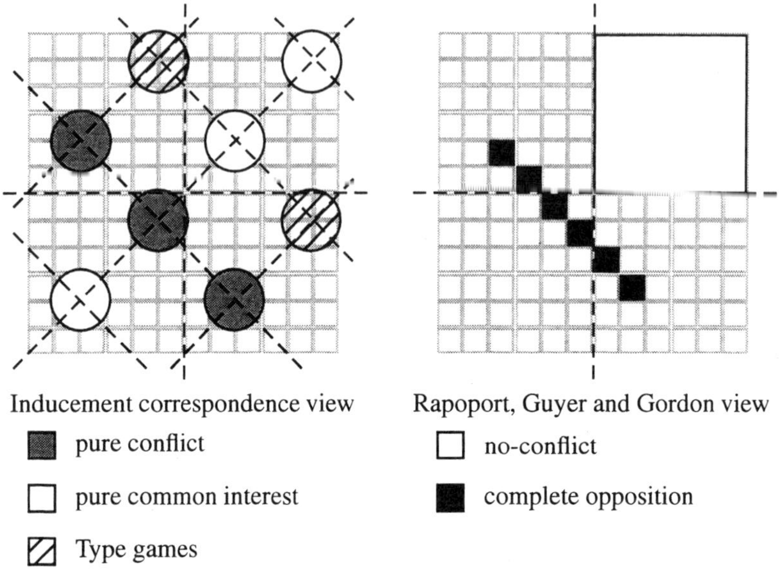


Figure 9.3: Locations of pure conflict, pure common-interest and Type games compared to Rapoport, Guyer and Gordon's no-conflict and complete opposition

have a distinct R&G reflection. If the table is folded along this axis each game will lie on the game produced from it by combining a reassignment $A$ and the reflection $R^{\nwarrow}$.

The quasi-symmetric games are shown in light grey. The $S'$ lines on Layers 2 and 4 mark another set of symmetry relationships previously described.

## 9.3 Conflict and common interest

The patterns of conflict and common interest are related to the symmetry axes. Symmetric and quasi-symmetric games mark the bands where individual players have strictly unmixed motives. As a result, the games at the intersections of the axes must be pure conflict, pure common-interest, or Type games, as we have defined them. At left in Figure 9.3 the pure conflict games are dark, pure cooperation white. Hatched circles mark the concentration of Type games.

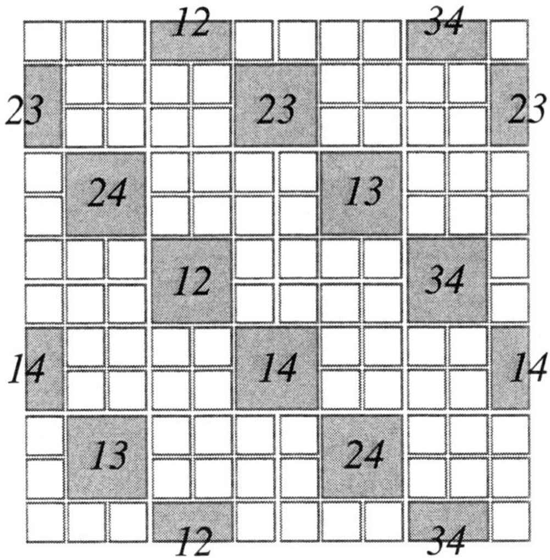


Figure 9.4: Linked hotspots

In Figure 9.3 contrast our pattern with the version proposed by Rapoport, Guyer and Gordon [24]. Their "no-conflict" games cover all of and only Layer 3 (where we have two out of three of our clusters of games of pure common interest), and their six "games of complete opposition" lie in a band that coincides with the three circles of pure conflict in the left panel.

## 9.4 Pipes and tiles

The Periodic Table contains information on the links between layers, making it possible to identify neighbours that are not adjacent in the table. Links can be treated as though they occur only at the centre of tiles. A 4-layer linked stack of tiles is a pipe. Hotspots are tightly linked pairs of tiles. In Figure 9.4, the grey areas along the quasi-symmetric axes represent hotspots and the white areas are pipes.

For the hotspots, common labels indicate the six pairs that are linked. Label numbering specifies which layers are involved. On the positive diagonal, for example, there are two tiles labelled "13" that form one hotspot. The first is on Layer 1, the second on Layer 3.

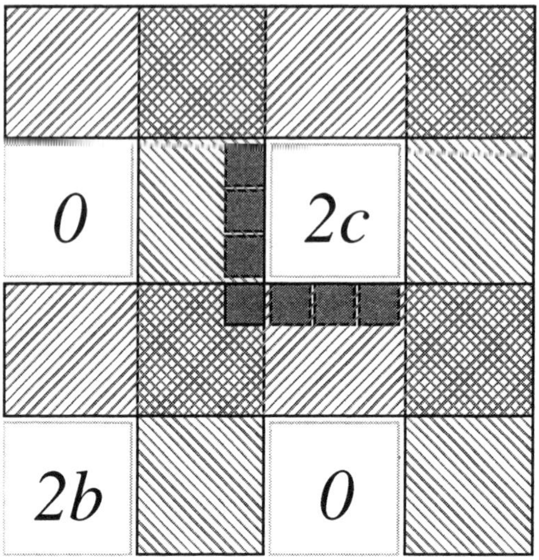


Figure 9.5: Dominant strategies and number of equilibria

The tiles in some hotspots and pipes appear split in Figure 9.4. The eight grey games in three areas labelled "14" constitute a single hotspot. Pipes may appear to split across layer boundaries as well[2]. Recall the PD pipe of Figure 9.1.

## 9.5 Two, one or no dominant strategies

Rapoport and Guyer [23] and Rapoport, Guyer and Gordon [24] cross-classify games according to several criteria, including whether the games "have two, one, or no dominating strategies" ([24] p. 19). In Figure 9.5 the games for which Row has a dominant strategy are marked with a positive cross-hatch and games for which Column has a dominant strategy are marked with a negative cross-hatch. Games that are double-hatched have dominant strategy equilibria.

Identifying games with dominant strategies divides the Periodic

[2]In the standard layout in Chapter 2 tiles are never split but no links between layers are directly shown.

Table into $3 \times 3$ squares. Each $3 \times 3$ square is composed of games with zero, one or two dominant strategies. It is clear that classification by dominance can be justified topologically.

### 9.5.1 Two, one or no Nash equilibria

Figure 9.5 also serves to classify games according to the number of equilibria [3]. All $2 \times 2$ games in which at least one player has a dominant strategy are dominance-solvable and always have a single Nash equilibrium. In the $3 \times 3$ regions with no dominant strategies the games have either two or no Nash equilibria. Games with two Nash equilibria are centred on the symmetric diagonal of the Periodic Table, and the regions with games with no Nash equilibria, labelled **0**, are cut by the quasi-symmetric diagonal[4].

Of the two equilibria games, the Battle of the Sexes set[5] on Layer 1 is labelled **2b** and the Coordination games on Layer 3 are labelled **2c**. It is worth noting that the Prisoner's Dilemma lies at the corner of region **2c** and, less obviously, **2b**.

#### Pareto-inefficiency

The only games with unique Pareto-inefficient equilibria are games in the Prisoner's Dilemma Family shown in dark grey on Figure 9.5. The games in **2c** have two Nash equilibria, one of which is Pareto-inefficient.

### 9.5.2 Dominant strategies and unmixed interests

Combining Figure 9.5 with the locations of the games with individually unmixed motives in Figure 9.3 to make Figure 9.6 reveals that

[3]Rapoport, Guyer and Gordon partially cross-classify games according to whether the "natural outcomes" are "equilibria, Pareto-optimal, both, or neither" ([24] p. 19). The natural outcome is a peculiar and essentially *ad hoc* construct designed to pick a single outcome in games with two equilibria.

[4]These are Rapoport and Guyer's Category 10, *Cycle games* [23].

[5]These are Rapoport and Guyer's Categories 8 and 9, *Pre-emption games* [23]. Category 8 is distinguished in having the maxi min solution differ from the Nash equilibria, a distinction that the authors ignore for other games.

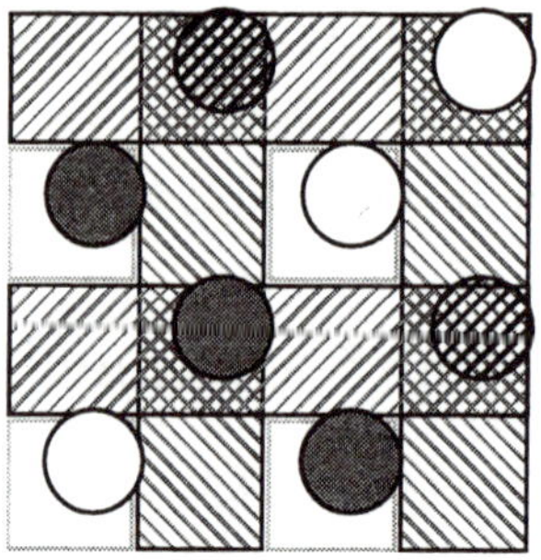

Figure 9.6: Dominant strategies and unmixed-motive games

many of the games with unmixed motives appear in the regions with dominant strategy equilibria. These are the games that Hamburger [11] called *separable*. The remainder are concentrated in the same relative position in the regions without dominant strategies.

## 9.6 Social dilemmas

At the centre of the Periodic Table lie the most interesting and problematic games of all, the Social Dilemmas. These are games with multiple and/or Pareto-inferior equilibria. Rational agents choosing independently are likely to produce inefficient outcomes.

The Periodic Table highlights the close connections between the Prisoner's Dilemma (black in Figure 9.7), the Alibi games (dark grey) and the Coordination games (light grey). In the lower left the Battle of the Sexes games are also light grey[6]. Coordination and Battle of the Sexes games form the 13 hotspot, outlined in black.

Seven of the 12 symmetric games are in this group: from the lower left they are Chicken, Battle of the Sexes games, the Prisoner's

[6]Some definitions of the social dilemmas are much more restrictive, including only variants of the Prisoner's Dilemma. We prefer a broader definition because it includes the entire well-defined and strongly interconnected region in which inferior outcomes are likely. A reasonable case can be made either for including the Battle of the Sexes games because they may yield Pareto-inferior outcomes, or for excluding them because they lack Pareto-inferior equilibria.

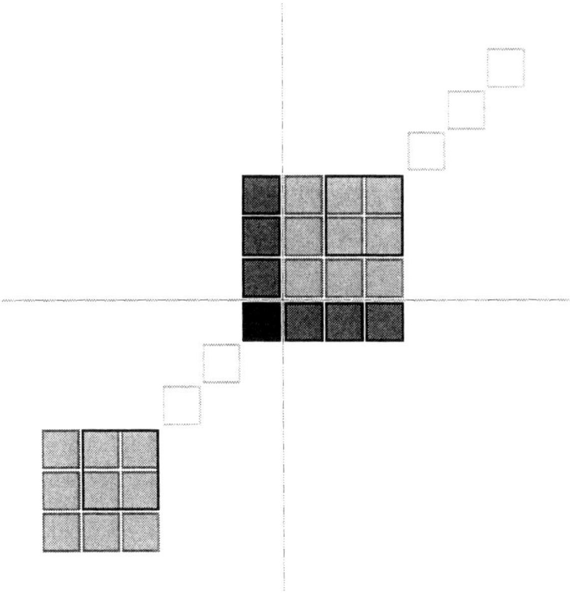

Figure 9.7: Dominant strategies and equilibria

Dilemma, Stag Hunt and two coordination games. In the light grey regions there are both symmetric and asymmetric games, repeating the pattern in the PD Family.

## 9.7 Previous typologies

The typology proposed by Rapoport and Guyer [23] in 1966 and developed further with Gordon [24] in 1976 is the only current alternative to the Periodic Table. It is really a multiple cross-classification scheme. Table 9.1 reconciles their numbering scheme and our own topologically-based indexing system. (It also provides the numbers used in Steven Brams' *Theory of Moves* [5] and lists common names for many of the games.)

With the exception of what they call "natural outcomes", the distinctions made in [23] and [24] are familiar and non-controversial. Both studies suggest that the categories should be understood as a hierarchical structure – the later work actually fits the games into a branching structure based on the classification of species.

1. **Phylum**- based on conflict or common interest
   - *N:* No-conflict games
   - *M:* Mixed-motive games
   - *Z:* Games of complete opposition
2. **Order** based on the number of players with dominant strategies
   - $D_0$*:* no dominant strategies
   - $D_1$*:* one dominant strategy
   - $D_2$*:* two dominant strategies
3. **Class** based on whether the "natural outcome" is a Nash equilibrium
   - *E:* The "natural outcome" is a Nash equilibrium
   - *e:* The "natural outcome" is not a Nash equilibrium
4. **Subclass** based on whether the "natural outcome" is Pareto-optimal
   - *P:* The "natural outcome" is Pareto-optimal
   - *p:* The "natural outcome" is Pareto-inefficient
5. **Genus** Eight genera based on a fragile hierarchy of stability notions.

The fanciful biological metaphor is no real help in understanding relationships among the games: there is simply no natural hierarchical order to these categories. In fact, a hierarchical classification tends to obscure important associations. The PD Family, for example, is divided into games with one or two dominant strategies before considering Pareto-efficiency or the Nash equilibria. The Prisoner's Dilemma ends up in a separate class of one because it has both a dominant strategy equilibrium and a deficient equilibrium. The Battle of the Sexes games are also split into two categories, while the Coordination games are submerged in the no-conflict Phylum.

There are some additional problems worth mentioning:

1. The Rapoport-Guyer-Gordon treatment of conflict and common interest is incomplete. Chapter 8 proposed a more systematic approach based on the topology of the $2 \times 2$ games.

2. Class and Subclass employ the concept of a "natural outcome" which does not apply to games without Nash equilibria and is defined in an *ad hoc* manner for games with two Nash equilibria. It is in our view more useful to identify regions with zero, one or two Nash equillibria. Furthermore, if it matters that the maxi-min solution concept does not select a Nash equilibrium among Battle of the Sexes games, then it is likely to matter in other cases as well. The version of the Periodic Table in this chapter identifies the Nash and maxi-min outcomes for all games.

3. The Rapoport-Guyer-Gordon classification by stability is not empirically based. It is at best a complex of hypotheses about behaviour. It is not clear, however, that behaviour even respects the boundaries of the ordinal games. In Chapter 10 we provide some evidence that behavior partitions the space of real-valued games rather differently from the way ordinality does.

# 9.8 A summary

The Periodic Table presented in this chapter captures the features that the Rapoport-Guyer-Gordon typology presents, and others. It is easier to use however, more flexible, and better designed for generating testable hypotheses about the $2 \times 2$ games. Like the Periodic Table of Elements, the Periodic Table of the $2 \times 2$ games reveals fundamental relationships that will guide research and make the subject more teachable.

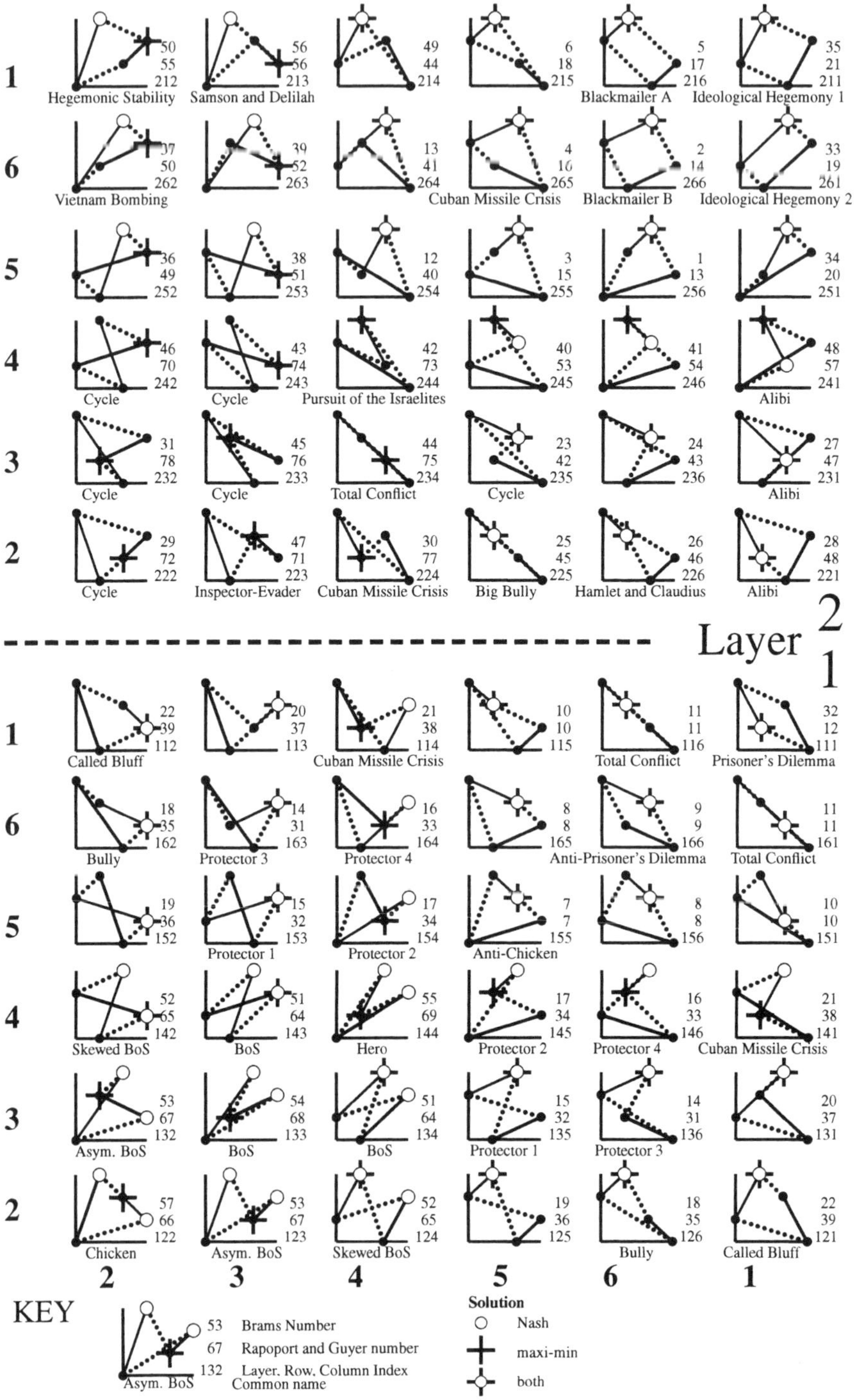
1
50 55 212 Hegemonic Stability
56 56 213 Samson and Delilah
49 44 214
6 18 215
5 17 216 Blackmailer A
35 21 211 Ideological Hegemony 1
6
37 50 262 Vietnam Bombing
39 52 263
13 41 264
4 16 265 Cuban Missile Crisis
2 14 266 Blackmailer B
33 19 261 Ideological Hegemony 2
5
36 49 252
38 51 253
12 40 254
3 15 255
1 13 256
34 20 251
4
46 70 242 Cycle
43 74 243 Cycle
42 73 244 Pursuit of the Israelites
40 53 245
41 54 246
48 57 241 Alibi
3
31 78 232 Cycle
45 76 233 Cycle
44 75 234 Total Conflict
23 42 235 Cycle
24 43 236
27 47 231 Alibi
2
29 72 222 Cycle
47 71 223 Inspector-Evader
30 77 224 Cuban Missile Crisis
25 45 225 Big Bully
26 46 226 Hamlet and Claudius
28 48 221 Alibi
Layer 2
Layer 1
1
22 39 112 Called Bluff
20 37 113
21 38 114 Cuban Missile Crisis
10 10 115
11 11 116 Total Conflict
32 12 111 Prisoner's Dilemma
6
18 35 162 Bully
14 31 163 Protector 3
16 33 164 Protector 4
8 8 165
9 9 166 Anti-Prisoner's Dilemma
11 11 161 Total Conflict
5
19 36 152
15 32 153 Protector 1
17 34 154 Protector 2
7 7 155 Anti-Chicken
8 8 156
10 10 151
4
52 65 142 Skewed BoS
51 64 143 BoS
55 69 144 Hero
17 34 145 Protector 2
16 33 146 Protector 4
21 38 141 Cuban Missile Crisis
3
53 67 132 Asym. BoS
54 68 133 BoS
51 64 134 BoS
15 32 135 Protector 1
14 31 136 Protector 3
20 37 131
2
57 66 122 Chicken
53 67 123 Asym. BoS
52 65 124 Skewed BoS
19 36 125
18 35 126 Bully
22 39 121 Called Bluff
2 3 4 5 6 1
KEY
53 Brams Number
67 Rapoport and Guyer number
132 Layer, Row, Column Index
Asym. BoS Common name
Solution
Nash
maxi-min
both

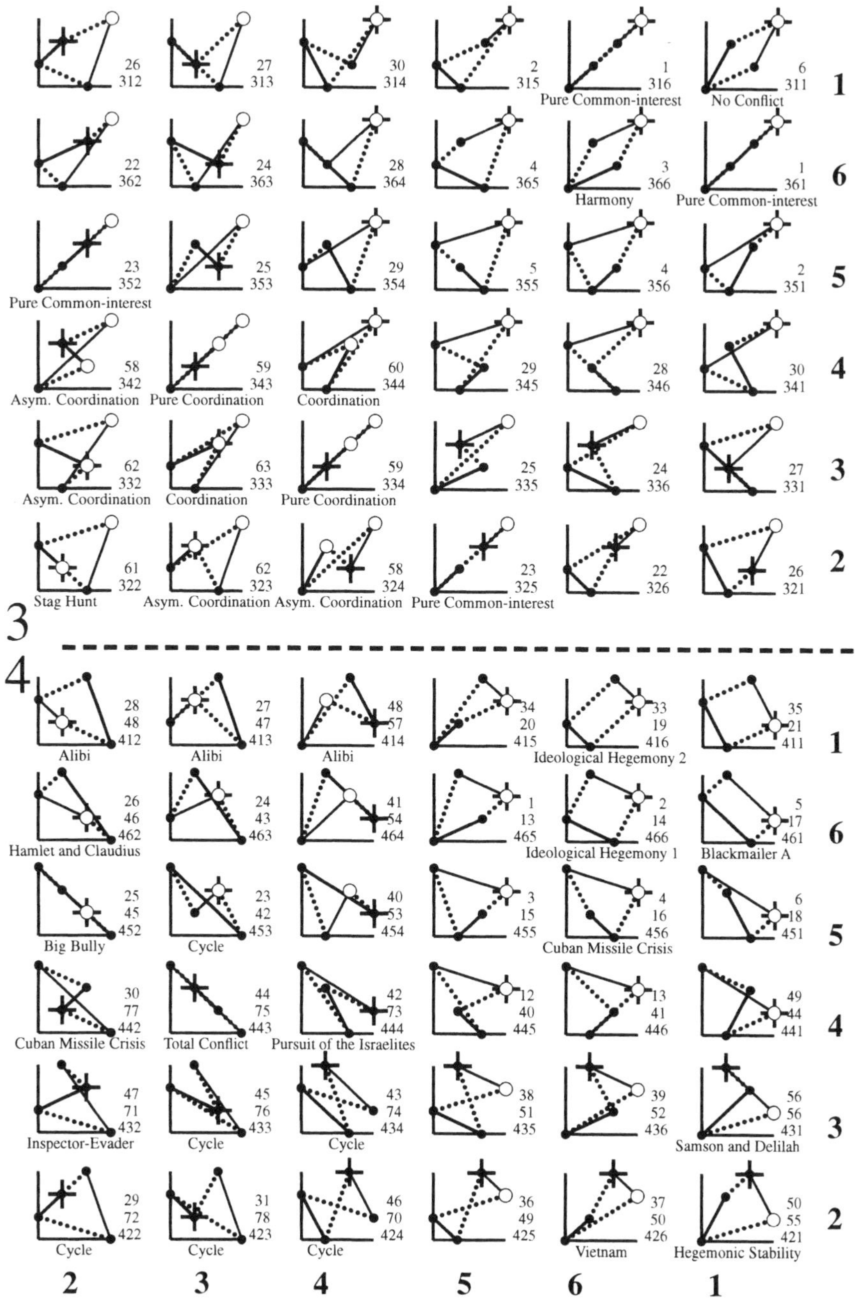


Figure 9.8: The complete Periodic Table of the $2 \times 2$ games

| index | $R \nwarrow A$ | R&G | Brams | Name |
|---|---|---|---|---|
| ***111*** | | 12 | 32 | Prisoner's dilemma, Deadlock |
| **121** | 112 | 39 | 22 | Polish Crisis, 1980-1, Union-Confederacy crisis, Called Bluff |
| ***122*** | | 66 | 57 | Chicken, Cuban Missile crisis, Brinksmanship, Pre-emption |
| **131** | 113 | 37 | 20 | |
| **132** | 123 | 67 | 53 | Asymmetric Battle of the Sexes, Pre-emption |
| ***133*** | | 68 | 54 | Battle of the Sexes, Benevolent Chicken, Pre-emption, Leader |
| **141** | 114 | 38 | 21 | Cuban missile crisis |
| 142 | **124** | 65 | 52 | Skewed Battle of the Sexes |
| 143 | **134** | 64 | 51 | Battle of the Sexes, Luke and Matthew |
| ***144*** | | 69 | 55 | Hero, Anti-Battle of the Sexes, Apology, Let-George-Do It, Pre-emption |
| 151 | **115** | 10 | 10 | |
| 152 | **125** | 36 | 19 | Type game |
| 153 | **135** | 32 | 15 | Protector 1 |
| 154 | **145** | 34 | 17 | Protector 2 |
| ***155*** | | 7 | 7 | Anti-Chicken |
| 161 | **116** | 11 | 11 | Total Conflict |
| 162 | **126** | 35 | 18 | Bully |
| 163 | **136** | 31 | 14 | Protector 3 |
| 164 | **146** | 33 | 16 | Protector 4 |
| **165** | 156 | 8 | 8 | |
| ***166*** | | 9 | 9 | Anti-Prisoner's Dilemma |
| **211** | 411 | 21 | 35 | Type game, Ideological Hegemony 1 |
| **212** | 421 | 55 | 50 | Vietnam Bombing, Iran hostage crisis, Hegemonic Stability |
| **213** | 431 | 56 | 56 | Samson and Delilah |
| **214** | 441 | 44 | 49 | |
| **215** | 451 | 18 | 6 | |
| **216** | 461 | 17 | 5 | Type game, Iran hostage crisis, Blackmailer A, Ideological Hegemony 2 |
| 221 | **412** | 48 | 28 | Alibi |
| **222** | 422 | 72 | 29 | Cycle |
| **223** | 432 | 71 | 47 | Cycle, Inspector-Evader |
| 224 | **442** | 77 | 30 | Cuban missile crisis, Cycle |
| 225 | **452** | 45 | 25 | Total Conflict, Big Bully |
| 226 | **462** | 46 | 26 | Hamlet and Claudius |
| 231 | **413** | 47 | 27 | Alibi |
| **232** | 423 | 78 | 31 | Cycle |
| **233** | 433 | 76 | 45 | Cycle |
| 234 | **443** | 75 | 44 | Total Conflict, Cycle |
| 235 | **453** | 42 | 23 | Cycle |
| 236 | **463** | 43 | 24 | |
| 241 | **414** | 57 | 48 | Alibi, Revelation |
| 242 | **424** | 70 | 46 | Cycle |
| 243 | **434** | 74 | 43 | Cycle |
| **244** | 444 | 73 | 42 | Pursuit of the Israelites, Cycle |
| 245 | **454** | 53 | 40 | |
| 246 | **464** | 54 | 41 | |
| **251** | 415 | 20 | 34 | |
| **252** | 425 | 49 | 36 | Type game |
| **253** | 435 | 51 | 38 | |
| **254** | 445 | 40 | 12 | |
| **255** | 455 | 15 | 3 | |
| **256** | 465 | 13 | 1 | |
| **261** | 416 | 19 | 33 | Type game, Ideological Hegemony 2, Blackmailer A |
| **262** | 426 | 50 | 37 | Vietnam bombing |
| **263** | 436 | 52 | 39 | |
| **264** | 446 | 41 | 13 | |
| **265** | 456 | 16 | 4 | Cuban missile crisis |
| 266 | **466** | 14 | 2 | Type game, Blackmailer B, Ideological Hegemony 1 |
| ***311*** | | 6 | | No Conflict |
| 321 | **312** | 26 | | |
| ***322*** | | 61 | | Stag Hunt |
| 331 | **313** | 27 | | |
| 332 | **323** | 62 | | Asymmetric coordination |
| ***333*** | | 63 | | Coordination |
| 341 | **314** | 30 | | |
| **342** | 324 | 58 | | Asymmetric coordination |
| **343** | 334 | 59 | | Pure common-interest coordination |
| ***344*** | | 60 | | Coordination |
| **351** | 315 | 2 | | |
| **352** | 325 | 23 | | Pure common-interest |
| **353** | 335 | 25 | | |
| **354** | 345 | 29 | | |
| 355 | | 5 | | |
| **361** | 316 | 1 | | Pure common-interest |
| **362** | 326 | 22 | | |
| **363** | 336 | 24 | | |
| **364** | 346 | 28 | | |
| 365 | **356** | 4 | | |
| ***366*** | | 3 | | Harmony, Anti-No Conflict |

Table 9.1: Game numbers from Rapoport and Guyer's (R&G) and Brams' typologies, with names from various sources

# Chapter 10

# The real world

In this chapter we present a series of evolutionary experiments designed to clarify the relationship between the topology of the ordinal games and a topology for real-valued $2 \times 2$ games.

Every ordinal game is associated with an equivalence class in the real-valued games. That equivalence class consists of all the games whose payoffs satisfy the ordering in the ordinal game.

The issues we are concerned with are straightforward: How well does the ordinal game represent its equivalence class? Do the boundaries of the equivalence classes correspond to the boundaries of interesting or important behaviours in the space of real-valued games?

The questions are important because experimental results are usually reported for specific payoff matrices. For example, the vast majority of simulation experiments with the Prisoner's Dilemma use a single payoff matrix, most often the one popularized by Axelrod in his tournaments [1][2]. If ordinally equivalent games are behaviourally equivalent, then studies conducted with a single payoff bi-matrix will provide useful explanations or predictions about the entire equivalence class. If not, then the study of games is a considerably more formidable task.

The experiments in this chapter show, first, that ordinal boundaries do not generally correspond to behavioural boundaries, and second, that the topological structure developed in this book can be profitably extended to organize the exploration of real spaces.

**NOTATION**

| | Defect | Cooperate |
|---|---|---|
| Coop | S, T | C, C |
| Defect | D, D | T, S |

**VALUES**

| | Defect | Cooperate |
|---|---|---|
| Coop | 0, 5 | 3, 3 |
| Defect | 1, 1 | 5, 0 |

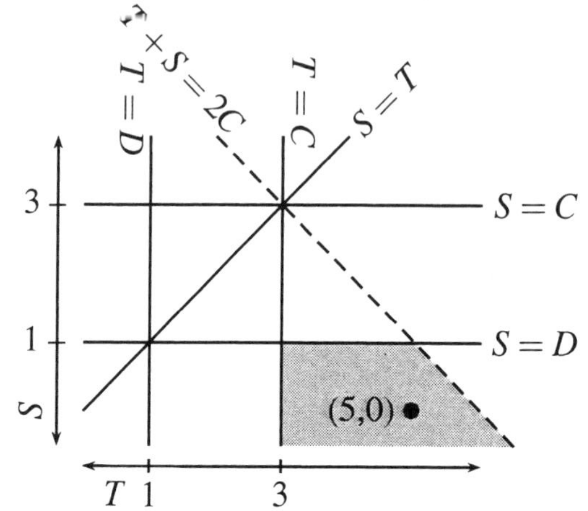


Figure 10.1: Real-valued symmetric games with the restricted Prisoner's Dilemma region in grey

## 10.1 A real-valued version of the model

Figure 10.1 embeds the model of the symmetric games from Chapter 4 in the real plane. Lines are labeled to reflect Axelrod's ([1], [2], [6]) familiar terminology. $C = 3$ is the "Cooperation" payoff earned if both cooperate, $D = 1$ the payoff if both "Defect". The axes, $T$ and $S$ are scaled in terms of $C$ and $D$[1].

Any point $(T,S)$ represents a symmetric $2 \times 2$ game, and any point in the lower right represents a Prisoner's Dilemma. The dashed line with negative slope represents a condition sometimes added to the definition of the Prisoner's Dilemma for real-valued games. The combinations of $S$ and $T$ on the line yield the same joint payoff as a common strategy of cooperation. In games above the line, $S+T$

[1]To locate any symmetric game $(c,d,t,s)$ on this plane, map $c$ to $C(=3$ *in this case*$)$, $d$ to $D(=1)$. Values of $S$ and $T$ are computed by linear transformations: $S = C + (s-c)\frac{C-D}{c-d}$ *and* $T = C + (t-c)\frac{C-D}{c-d}$.

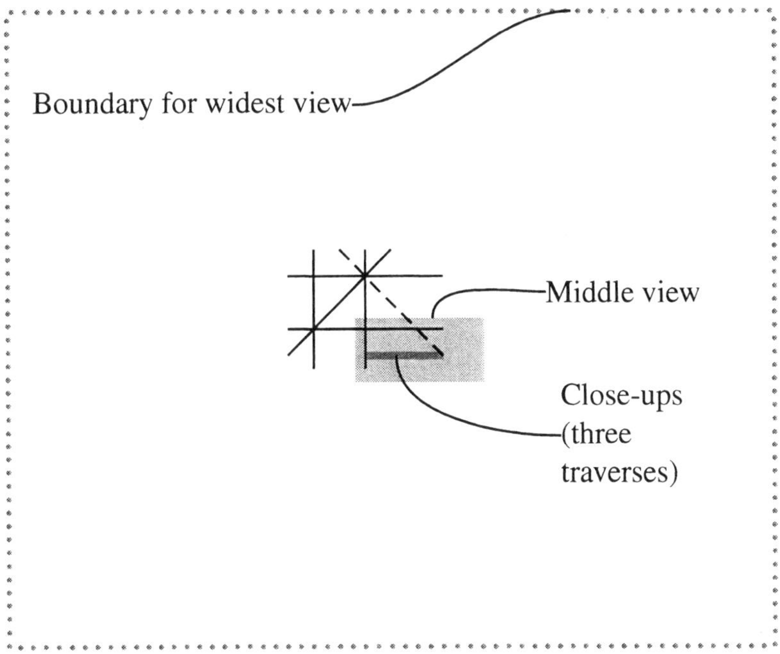


Figure 10.2: Territories explored in evolution experiments

yields a higher joint payoff than cooperation, and in games below the line, a lower joint payoff. The restricted Prisoner's Dilemma region is shown in grey in Figure 10.1.

Axelrod's experiments were conducted with the game $(T,S) = (5,0)$. As a result, $(T,C,D,S,) = (5,3,1,0)$ has become the *de facto* standard payoff set for Prisoner's Dilemma. $T$ is the payoff for Row if Row succumbs to "Temptation" (Defects) and $S$ is the "Sucker" payoff that Column receives if she continues to Coooperate when Row Defects.

Any game in the Prisoner's Dilemma region can be transformed into Chicken then into the Battles-of-the-Sexes by increasing the Sucker payoff $S$. Alternately, a Prisoner's Dilemma game can be transformed to Stag Hunt and the Coordination games by decreasing the Temptation payoff.

| Strategy | key | description |
|---|---|---|
| *All_C* |  | always cooperate |
| *All_D* | | always defect |
| *TFT* |  | cooperate first, then match opponent's last move |
| *aTFT* | | defect first, then *invert* opponent's last move |
| *C_Alt* |  | cooperate first, then alternate |
| *D_Alt* | | defect first, then alternate |

Table 10.1: Strategy key for the evolutionary game tournaments

## 10.2 An evolutionary investigation

The evolutionary methodology is well known [2] [28] [29]. Members of a population of strategies compete one-on-one against all others in a repeated game. The proportion of each strategy in the population is revised after each round of play based on the relative total score in that round. The process is repeated until the population stabilizes.

To investigate the influence of the payoffs on the final population we run the same experiment varying the $S$ and $T$ payoffs systematically. We then compare the population profiles.

Our first experiment covers the entire region bounded by the dotted rectangle in Figure 10.2. It includes samples of all twelve games. We then sample the smaller, light grey region more densely. The final results sample repeatedly along three lines that cross the Prisoner's Dilemma region. One of the traverses includes the classic case.

The strategies are described in Table 10.1.

Each strategy plays all strategies in a 200-round repeated game. The population evolves through 1000 generations. With the classic payoffs, the final population profile is over 99% *TFT* and less that 1% *All_C*.

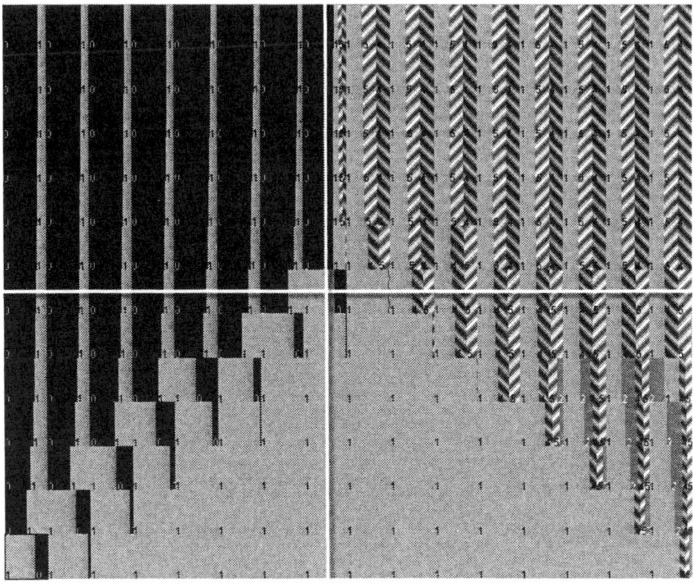

Figure 10.3: Wide view: population profiles over all 12 games

**A very wide view**

Figure 10.3 presents the final population mix at each point on a grid from $S = -11$ to $S = 13$ and $T = -11$ to $T = 19$, the dotted rectangle in Figure 10.2. The sample spacing is 2. White lines at $T = C$ and $S = D$ mark the quadrants in Figure 10.1.

In this composite diagram, the population profile for each sample is shown as a horizontal stacked-bar graph centred at the coordinates $(s,t)$. For example, the fraction of each rectangle that is black indicates the portion of the survivors playing *All_C*. The overall effect is to show how the success of each strategy varies across the payoff plane.

Figure 10.3 is an *ecological map*. It shows which species – in our case the strategies are species – thrive in regions that reward certain joint behaviours differently.

Broadly speaking there are three behaviourally distinct regions at this scale. *TFT* dominates roughly in Prisoner's Dilemma, Stag

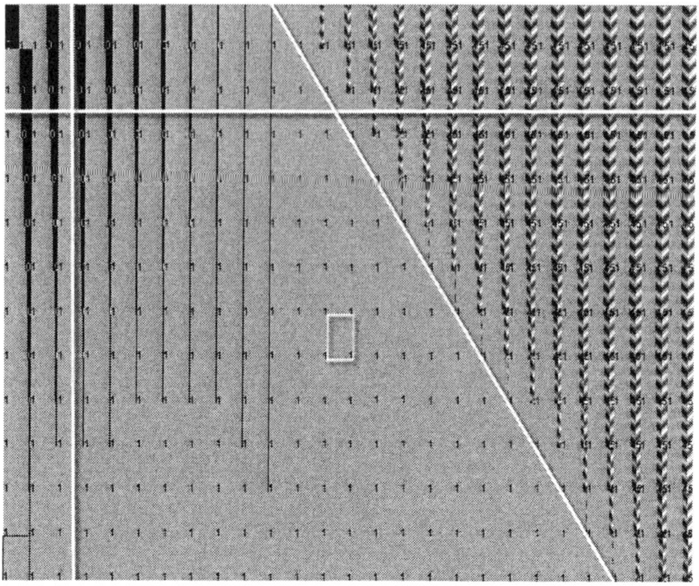

Figure 10.4: Close-up: population profiles for the Prisoner's Dilemma region

Hunt and the standard Coordination game. *All_C* dominates from Anti-Coordination all the way to the boundary of the Anti-Battle of the Sexes. In the Anti-Battle of the Sexes through Chicken, *TFT* and a mixture of the alternating strategies prosper.

### A closer view

Figure 10.3 shows what happens with large values of $T$ and $S$. For values near $C$ and $D$ we "zoom in", producing Figure 10.4. The range $2.6 \leq T \leq 7.6$ and $-1.0 \leq S \leq 1.4$ corresponds to the smaller light grey area in Figures 10.2. The step size is now 0.2. The diagonal line is where $T + S = 2C$.

Below the line is the restricted PD territory (bounded by Chicken above and Stag Hunt to the left). Above the line taking turns yields the largest joint payoff for repeated play. These games are ordinally Prisoner's Dilemmas, but are economically distinct with real-valued payoffs.

Clearly *TFT* dominates in the restricted Prisoner's Dilemma region, and for the part of the Chicken region directly above. To the right of the diagonal line *TFT* shares the territory with *C_Alt* and *D_Alt*. In the upper left *All_C* intrudes.

At the resolution of this diagram, the $< 1\%$ of *All_C* does not even show up for the $(5,0)$ rectangle (outlined in white). On the basis of this map, it is difficult to argue that the results of a classic Prisoner's Dilemma experiment should not be extrapolated to the entire (restricted) Prisoner's Dilemma region.

### 10.2.1 The ecology of errors

When the possibility of error is introduced, the situation changes. An error occurs when a player's intended move is randomly switched to the opposite move. With errors the ecological boundaries shift significantly.

The diagrams of Figure 10.3 and 10.4 are repeated at top left of Figures 10.5 and 10.6 respectively. The remaining three panels of each figure show the results of rerunning the experiments with increasing error rates in clockwise order. At the top right approximately one in $10,000$ moves is corrupted, at the bottom right, one in $1,000$, and at the bottom left, one in 100.

#### Results

- With one error in 10,000, the broad view shows instability in the Coordination game region, lower left. In the close-up, the *PD* region is a field of chaotic encounter between *All_C* and *TFT* where random factors propel one or the other to dominance[2]. It appears that the evolutionary process is unstable in part of the Prisoner's Dilemma region.

- When the error rate rises to one in 1,000 (lower right map in each figure), *TFT* establishes itself in the dominance-solvable games of the upper left quadrant while *All_C* continues to push further across the Prisoner's Dilemma region. However, the

[2]These images present single runs of each evolutionary contest to emphasize the instability with errors in play.

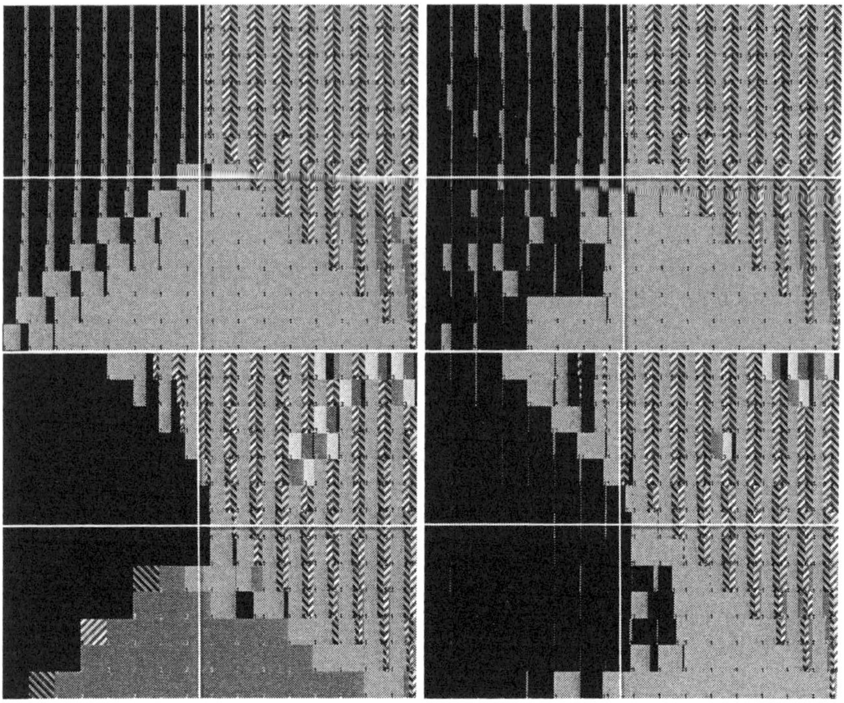

Figure 10.5: Four wide views with errors. Clockwise from the top left the error rates are 0, 0.0001, 0.001, 0.01

closeup shows that for values of $T$ and $S$ near $C$ and $D$ the situation in the restricted PD region is very unstable. Even $All\_D$ and $aTFT$ can emerge as the dominant species.

- Finally, with an error rate of one in 100 (lower left), $All\_D$ has become established in the area with low values of $S$. In the detailed view, the chaotic behaviour in the left side of Prisoner's Dilemma continues but the right side nearer the diagonal has apparently stabilized with a mixture of the alternating strategies and $TFT$.

The bottom left image in Figure 10.6 provides the most striking evidence of the sensitivity of populations to small displacements across the payoff plane unrelated to ordinal boundaries. The PD region exhibits two distinct patterns: where $T$ is near to $C$, the populations appear very unstable with four of the strategies each coming to

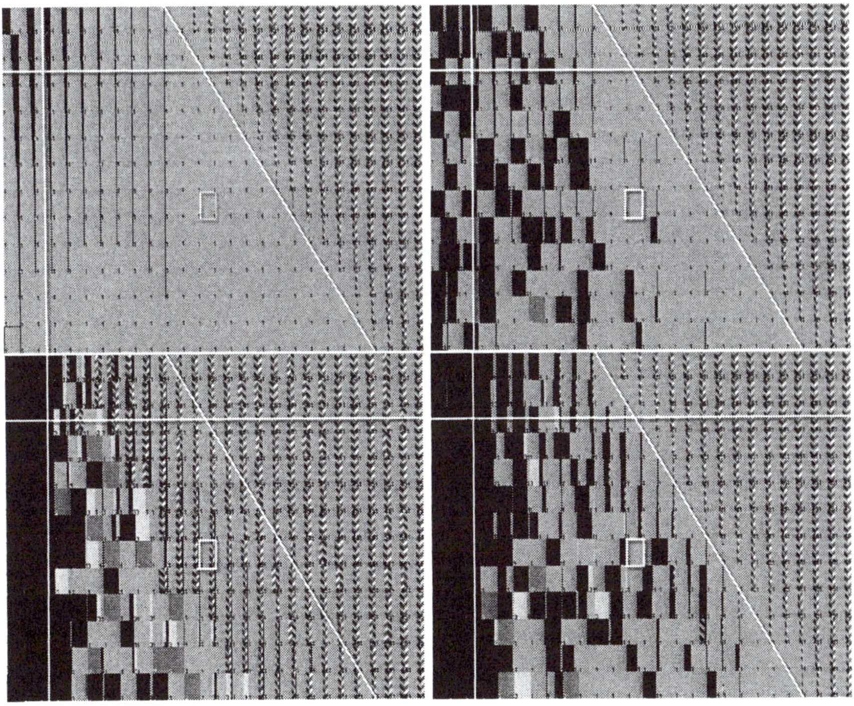

Figure 10.6: Effect of error in the Prisoner's Dilemma region

dominate in some tournaments; where $T$ approaches the boundary $S+T=2C$, the populations are stable with combinations of $TFT$, $C\_Alt$ and $D\_Alt$ coming to equilibrium in spite of the high error rate.

**Chaos crossing the PD region**

Figure 10.7 shows more detail along three cuts through the Prisoner's Dilemma territory from boundary to boundary. The value of $T$ varies from 3.00 to 6.00 with stepsize 0.05. $S$ takes three different values, 0.00, 0.02 and 0.04. Vertical stacked bars on each interval show the population proportion for each strategy in the last round. (Axelrod's case, the game defined by $(5,0)$, is identified in the lowest strip by a white rectangle.)

The wide grey bands on the right show $TFT$ making up roughly half of the population and combinations of $D\_Alt$ and $C\_Alt$ making up the rest. This population profile is stable from the right boundary

Figure 10.7: Crossing to chaos in the Prisoner's Dilemma region, $m = 0.01$

($T = 6$) to a point (between 4.1 and 4.5 depending on the row) where the behaviour becomes abruptly erratic.

The left side of all three traverses has the characteristic appearance of a chaotic region where randomness can drive the population to any one of many attractors. If an experiment had been conducted only with the classic payoff matrix (white rectangle), the chaotic behaviour would have been missed.

## 10.3 In conclusion: ordinal boundaries and real behaviour

These experiments have strong implications. Ordinal boundaries do not exactly correspond to behavioural boundaries. In some cases the boundary is approximately correct, in some cases the boundary is blurred, and in some cases behavioural boundaries cross ordinal regions. We conclude that:

> *The discrete order topology of the ordinal games [25] is important to understanding the relationship among $2 \times 2$ games but is insufficient for describing and predicting patterns of behaviour.*

When an experiment is conducted using particular payoffs, it is not justifiable to claim a result for an ordinal class of games. Only by systematically sampling the space around the original payoffs is it possible to ascertain how generally a result applies. It may apply to an entire ordinal class or it may not; it may also apply to other games identified as neighbours in the Periodic Table.

What we can say, and it is not a small point, is that is that the topological structure of the ordinal $2 \times 2$ games can be usefully extended to the real-valued games. That extension provides a systematic approach for teaching and suggests an agenda for research. The $2 \times 2$ gamess are no longer just a string of unconnected anecdotes.

# Glossary

A

**Actions (or Moves)** Alternatives available to a player at a specific point in the play of a game. In strategic form players do not have alternating moves, but it is common to discuss games with moves using the strategic form.

**Alibi games** Asymmetric members of the Prisoner's Dilemma Family. Dominance-solvable games with one Pareto-inefficient equilibrium. Games $g_{221}$, $g_{231}$, $g_{241}$, $g_{412}$, $g_{413}$, $g_{414}$.

**Anti-game** A game in which the roles of the players have been reversed. Each row of the payoff matrix is converted into a column and vice versa. Inducement correspondences have been reassigned.

**Assignment** Naming the players that choose the rows or columns. Identifying the player associated with each inducement correspondence.

**Assurance game** A generic name for the game more commonly known as *Stag Hunt* or *Stag and Hare*.

**Asymmetric information** Players have different information sets.

**Axelrod's tournament** A famous simulation of a repeated-PD tournament in the early 1980s in which strategies were played against each other to determine which would be most successful against all comers.

B

**Bargaining game** A two-player game in which players must both agree to one set of feasible outcomes or else accept a predetermined "disagreement" outcome.

**Battle of the Sexes (BoS)** Games with two equally efficient but distributionally different Nash equilibria. The BoS games allow for mutual gain with distributional conflict. A model of specialisation and possibly of social roles. Games $g_{123}$, $g_{124}$, $g_{132}$, $g_{133}$, $g_{134}$, $g_{142}$, $g_{143}$, $g_{144}$.

**Battle of the Two Cultures** An alternative story for the *Battle of the Sexes*.

**Best response/best reply** Player $i$'s best response to the *strategy profile* of all players except $i$, is the strategy that yields her the greatest payoff.

**Best response analysis** An approach to analysing games that focuses on the best response for each player to all the moves of the other player.

**Bi-matrix** A convenient matrix version of the payoff function showing payoffs for both players in a single cell.

C

**Chicken** A $2 \times 2$ symmetric game similar to the PD, except the "sucker payoff" is not worse than the consequence of mutual defection. As a result, mutual defection is not an equilibrium, and there are two distributionally distinct Nash equilibria. Game $g_{122}$.

**Column (of game layout)** A set of 6 ordinal games closed under the operations $(R_{12}, R_{23})$. There are 24 columns. In each column, all games share the same *layer index* $\ell$ and the same *column index c*.

**Column index** $c$ The third number in the ordinal $2 \times 2$ games number system. $c \in \{1,2,3,4,5,6\}$.

**Column swap** $C_{ij}$ reverses the ranking of the outcomes ranked $i$ and $j$ for the column player. $|i-j| = 1$. A column swap is *not* equivalent to a column exchange on a matrix.

**Common interest, game of** Game in which moves that benefit one player benefit all players. Defined here as a game in which the inducement correspondences are positively sloped. Not equivalent to *"no-conflict"* games in Rapoport and Guyer, or Brams, or to Schelling's "pure-collaboration" game.

**Common knowledge** Information that all players have and that all players know the other players have.

**Complete information** A player has complete information if she knows who all the players are, all the actions available to each player, and all the potential outcomes for each player. Otherwise, she has incomplete information. Games in strategic form have complete information.

**Constant-sum game** A game in which the sum of the payoffs for all players is a constant. Zero-sum games are a special case. In a constant-sum game there is conflict over distribution, and players do not generally have a common interest. There may be a Nash equilibrium. Special case of *constant rank-sum* ordinal game.

**Constant rank-sum game** A game of pure conflict in which the sum of ranks for all players for every strategy combination is constant. The ordinal analogue of *constant-sum game*. Games $g_{161}$, $g_{452}$, $g_{443}$, $g_{234}$, $g_{225}$, $g_{116}$.

**Coordination game** A social situation with interdependent decisions, a coincidence of interests, and at least two proper coordination equilibria. A game with more than one Nash equilibrium in which no player would be better off if any one player (including him/herself) unilaterally made a different choice. Games $g_{323}$, $g_{324}$, $g_{332}$, $g_{333}$, $g_{334}$, $g_{342}$, $g_{343}$, $g_{344}$.

D

**Defect** To select an action which must reduce the payoff for other players more than it increases one's own. The notion of defection is essentially defined within the payoff structure of the Prisoner's Dilemma for the theory of games.

**Distributional conflict** Conflict arising because a gain for one individual or group is associated with a loss for another.

**Dominance layout** A display of the $2 \times 2$ games in four layers with each layer configured so the row player has a dominant strategy in the bottom three rows and the column player has a dominant strategy in the left three columns. The game $g_{\ell 11}$ is in the third row and column of each layer.

**Dominance solvable (DS) games** The class of games for which a unique outcome can be selected by sucessive elimination of dominated strategies. If either player in a $2 \times 2$ game has a dominant strategy the game is dominance solvable. Solutions found this way are always Nash equilibria. Three quarters of $2 \times 2$ games are dominance solvable.

**Dominant strategy** A strategy which is a best response to every strategy combination of the other players. Dominant strategies do not always exist.

**Dominant strategy equilibrium** An equilibrium in which each player plays her dominant strategy. One quarter of $2 \times 2$ games have dominant strategy equilibria.

**Dominated strategy** A strategy which is not a best response to any combination of strategies chosen by other players.

E

**Efficiency** The absence of waste or unused resources. Not well defined for ordinal games. See *Pareto-efficiency*.

**Equilibrium concept** A plausible principle for identifying outcomes of interest as potential behavioural outcomes, generally based on stability considerations. See *Solution concept*. For examples see *Dominant strategy equilibrium* and *Nash equilibrium*.

**Elimination of dominated strategies** A principle for strategy selection by elimination of strategies that are not a best response to any combination of strategies chosen by other players. Applied iteratively it may yield a *Nash equilibrium*.

**Externalities** Effects that fall on someone other than the decision-maker. Negative externalities are inherent in any negatively sloped inducement correspondence. Positive externalities may be described with positively sloped inducement correspondences.

**Euler number** ($\chi$) Relates the number of vertices V, faces F, and edges E, of a polyhedron. $\chi \equiv V - E + F = \chi(g)$, where $g$ is the the closed surface the graph is embedded in.

F

**Face** The intersection of an n-dimensional polytope with a tangent hyporplane. Zero-dimensional faces are known as polyhedron vertices (nodes), one-dimensional faces as polyhedron edges.

**Fairness** Term describing either processes or outcomes which appeal because they incorporate equality or recognized entitlements.

**Focal point (Schelling point)** An outcome with a property that distinguishes it from other candidate solutions. May serve to coordinate play in a coordination game. A focal point has salience (i.e., it stands out), as for example the $(4,4)$ outcome in a Coordination game.

**Form** A representation of a situation containing all the information that may be used to predict or prescribe an outcome.

G

**Game** A situation of strategic choice in which payoffs are dependent on the decisions of others as well as one's own decisions.

**Game theory** An approach to analysing social situations. Often described as a "mathematical" theory and almost always associated with rational choice.

**Generators, Set of** A set of elements of a discrete group $G$ with the property that every element of $G$ can be expressed as a finite product of powers of the members of the set.

**Graph** A set of nodes, or vertices, and the set of edges that connect vertices in pairs. The graph of the $2 \times 2$ games contains 144 (game) nodes and 432 (swap) edges connecting them.

HIJK

**Hawk Dove game** A model from evolutionary game theory using the payoffs for *Chicken*. Game $g_{122}$.

**Hotspot** A set of eight games closed under the operations ($R_{12}$, $C_{12}$, $R_{34}$, $C_{34}$). There are six hotspots. The most connected subspaces in the space of the $2 \times 2$ games.

**Inducement correspondence** A mapping that identifies for each available action the set of outcomes the other player can reach. Developed by Joseph Greenberg in *A Theory of Social Situations*.

**Inducement correspondence for the Nash situation** The specific version of the *inducement correspondence* used in the order graph. "Row's inducement correspondence" is the set of outcomes induced *for* the row player (i.e. available to choose from) *by* the column player in selecting a specific strategy.

**Iterated dominance** An approach to selecting outcomes by repeatedly eliminating dominated strategies, first for one player then another.

L

**Layer (of game layout)** A set of 36 ordinal games closed under the operations $(R_{12}, C_{12}, R_{23}, C_{23})$. There are four layers. In a layer all games share the same *layer index* $\ell$.

**Layer index** $\ell$ The first number in the ordinal $2 \times 2$ games number system. $\ell \in \{1,2,3,4\}$.

M

**Maxi-min strategy** A strategy for which the worst possible payoff is at least as good as the worst payoff from any other strategy. The maxi-min payoff is the highest payoff that a player can guarantee herself. In the strict ordinal $2 \times 2$ games the outcome when both players use the maxi-min strategy can be found by eliminating the row with a 1 for the row player and the column with a 1 for the column player.

**Mixed interest, game of** A game in which at least one inducement correspondence is negatively sloped and at least one positively sloped. See *pure common-interest, pure conflict*, Chapter 8.

**Mixed strategy** A vector of values $p_i, \sum_i p_i = 1$, each representing the probability that a player will play strategy $i$. A mixed strategy Nash equilibrium is not defined for ordinal games.

**Moves** *Actions.*

**Multiple equilibria** The situation in which a game has more than one Nash equilibrium, as is the case for $\frac{1}{8}$ of the $2 \times 2$ games. When a game has multiple equilibria, game theory does not have a widely accepted criterion for selecting among them.

N

**Nash equilibrium** A strategy combination consisting of a best response for each of the players in the game to the choices of the others. At a Nash equilibrium, no player has an incentive to change her strategy unilaterally.

**No-conflict game** (*per* Rapoport and Guyer) A game in which the best outcome for both players is simultaneously feasible. All games on Layer 3.

**Normal form** *Strategic form.*

O

**Opposed interest, game of** Games in which moves that benefit one player harm others. Defined here as a game in which the inducement correspondences are negatively sloped.

**Order graph** A representation of the *strategic form* of a 2-person ordinal game in payoff space.

**Ordinal game** A game in which the payoffs can be ranked – best, second best, ... – but not added or subtracted. (See *Strict ordinal game*.)

P

**Pareto coordination game** Coordination game in which one Nash equilibrium Pareto-dominates others. All nine strict ordinal Coordination games are of this sort. Games $g_{322}$, $g_{323}$, $g_{324}$, $g_{332}$, $g_{333}$, $g_{334}$, $g_{342}$, $g_{343}$, $g_{344}$.

**Pareto-dominated outcome** An outcome in which the payoff for at least one player is lower than the payoff for that player in another feasible outcome and the payoff for no player is higher. In the Prisoner's Dilemma Family the Nash equilibrium is Pareto-dominated.

**Pareto-efficiency** Absence of an alternative feasible outcome preferred by all players.

**Payoff** The outcome of a game for a specific player should a specific strategy combination occur.

**Payoff function** A rule, or vector function, that associates a payoff vector with every possible strategy combination.

**Payoff matrix** A representation of the strategic form of a 2-person game in strategy space as a matrix of payoffs.

**Periodic Table of** $2 \times 2$ **games** A $12 \times 12$ layout of the 144 ordinal $2 \times 2$ games in which each game is located adjacent to games that are similar based on preference.

**Pipe** A set of sixteen games closed under the operations ($R_{12}$, $C_{12}$, $R_{34}$, $C_{34}$). There are six pipes.

**Player** An individual who makes decisions in a game.

**Player set** The set of players for a game.

**Preferences** The values of players that allow them to determine which outcome is preferred to which. Payoffs are only meaningful in terms of underlying preferences.

**Prisoner's Dilemma** The most famous game. Both players have dominant strategies and the equilibrium is Pareto-dominated by another feasible payoff combination. Game $g_{111}$.

**Prisoner's Dilemma Family** All games with unique Nash equilibria Pareto-dominated by other feasible payoff combinations which are not Nash equilibria. *Prisoner's Dilemma* and the *Alibi games.*

**Pure common-interest, game of** A game in which every payoff-improving strategy change for one player increases the payoffs for the other players.

**Pure conflict, game of** A game in which every payoff-improving strategy change for one player reduces the payoffs for the other players. See *Total conflict.*

QR

**Quasi-symmetric game** A game for which the order graph is symmetric around the positive and negative diagonals. (invariant under $R^{\nwarrow}$ and $R^{\swarrow}$) A quasi-symmetric game is not symmetric in the game-theoretic sense; it is symmetric under an exchange of payoffs between players at every outcome.

**Rational** Choosing that which is preferred. Acting to achieve what is preferred.

**Rational behaviour** Behaviour that can be explained as the result of rational choices.

**Rational set** In a bargaining game, the subset of possible outcomes for which no player is worse off than in the disagreement outcome. It would be irrational for any player to agree to a solution outside this set.

**Repeated games** When the players know that they will play the same game against the same players in the future, the game is a repeated game. The fact that the game will be repeated may affect the decisions of the players as they seek to establish a reputation or to signal their intentions to others.

**Row (of game layout)** A set of 6 ordinal games closed under the operations ($C_{12}$, $C_{23}$). There are 24 rows. In each row, all games share the same *layer index* $\ell$ and the same *row index* $r$.

**Row index** $r$ The second number in the ordinal $2 \times 2$ games number system. $r \in \{1,2,3,4,5,6\}$.

**Row swap** $R_{ij}$ reverses the ranking of the outcomes ranked $i$ and $j$ for the row player. $|i-j| = 1$. A row swap is not equivalent to an exchange of rows in a matrix.

S

**Separable game** A game in which the difference in payoff from changing strategy is independent of the other players' choices. Identified by Hamburger.

**Slice (of game layout)** A set of 24 ordinal games closed under the operations ($R_{12}$, $R_{23}$, $R_{34}$) or ($C_{12}$, $C_{23}$, $C_{34}$). There are 12 slices. In six slices, all games share the same *row index* $r$. In the other six, all games share the same *column index* $c$.

**Solution concept** A rule that selects a specific strategy combination or set of strategy combinations based on the structure of the game. Only a few solution concepts are widely accepted, the two best known being the *dominant strategy equilibrium* and the *Nash equilibrium*

**Social dilemma** A game in which rational individual choice does not reliably lead to an attractive outcome. The class includes the *Prisoner's Dilemma Family*, *Coordination games*, *Chicken* and possibly variants of the *Battle of the Sexes*.

**Stack (in a game layout)** A subspace of four ordinal games. There are 36 stacks. In each stack, all games share the same row index $r$ and the same column index $c$.

**Stag Hunt, Stag and Hare** Jean Jacques Rousseau described the situation in which two hunters can either jointly hunt a stag or individually hunt a hare. Hunting stags is quite challenging and requires cooperation. If either hunts a stag alone, the chance of success is minimal. Hunting stags is most beneficial for society but requires a lot of trust among its members. Game $g_{322}$.

**Standard layout** A display of the $2 \times 2$ games in four layers with the game $g_{\ell 11}$ is at the bottom left corner of each layer.

**Strategic form (Normal form)** The most familiar game form, consisting of a player set, $N$, a strategy set for each player, $\{S_i\}$, and a payoff vector-function from the strategy space to the utilities of the players. Often presented in a *payoff matrix* or *order graph*.

**Strategy (for player i)** (i) One of a set of available actions in a *strategy set*; $s_i \in S_i$. (ii) A complete plan of actions to be taken at all decision points in a game. A set of contingency plans.

**Strategy set (for player i)** The set $S_i$ of all strategies available to player $i$.

**Strategy combination/profile** A list of one strategy for each of the players in the game. Any one element $s = (s_1, s_2, .....s_n)$ of a strategy space $S$.

**Strategy space** The set of all the possible combinations of strategies for a game: $S = \times\{S_i\}$.

**Strict ordinal game** An ordinal game in which no outcomes have equal rank for any player.

**Sucker payoff** The lowest payoff in a Prisoner's Dilemma.

**Surgery** Topological thought-experiments which involve cutting and grafting surfaces.

**Swap operation** A minimal transformation applied to an ordinal $2 \times 2$ game by exchanging two consecutive payoffs for one player. There are six swap operators: $R_{12}, R_{23}, R_{34}, C_{12}, C_{23}, C_{34}$.

**Symmetric game** A game which is the same for both players. A game which is invariant under $R\nwarrow A$. There are 12 strict ordinal symmetric games.

**Symmetric information** When players all have the same information.

**Symmetric swap operation** A transformation applied to an ordinal $2 \times 2$ game by exchanging the same two consecutive payoffs for both players. There are three swap operators: $S_{12}$, $S_{23}$, $S_{34}$.

**Symmetry** An intrinsic property of a mathematical object which causes it to remain invariant under certain classes of transformations (such as rotation, reflection, inversion, or more abstract operations). The mathematical study of symmetry is systematized and formalized in the extremely powerful and beautiful area of mathematics called group theory.

T

**Temptation payoff** The highest available payoff in the Prisoner's Dilemma.

**Tile** A set of four games closed under the operations $(R_{12}, C_{12})$. There are 36 tiles in the space of $2 \times 2$ games.

**Topology** Mathematical study of properties of objects which are preserved through deformations by twisting and stretching. (Tearing and cutting are not allowed.)

**Torus** A surface possessing a single "hole". The usual torus in three-dimensional space is shaped like a doughnut. In general, the torus can have multiple holes, with the term $n$–torus used for a torus with $n$ holes.

**Total conflict, game of** Often identified with *constant sum games*. Any of 14 ordinal $2 \times 2$ games with all inducement correspondences negatively sloped.

**Tragedy of the commons** A multi-person Prisoner's Dilemma described by Hardin.

**Type game** A $2 \times 2$ game in which one player's self-interested choice always makes the other player better off but the other's self-interested choice always makes the first player worse off.

UVWXYZ

**Villarceau circles** Circles on the surface of a torus that go both around and through the hole. There are two distinct Villarceau circles which meet in two distinct points.

**Wiring** Any cycle joining the four outcomes on the order graph of a $2 \times 2$ game in payoff space.

**Zero-sum game** A game in which the sum of payoffs for all players for every strategy combination is zero. A zero sum game is a game of pure conflict. Analogous to a constant-sum game and a *constant rank-sum game*.

# Bibliography


[1] Axelrod, Robert. "Effective Choice in the Prisoner's Dilemma" in *Journal of Conflict Resolution, Vol. 24, No. 1*: 3 - 25. 1980.

[2] Axelrod, Robert. "More Effective Choice in the Prisoner's Dilemma" in *Journal of Conflict Resolution, Vol. 24, No. 3*: 379 - 403. 1980.

[3] Beaufils, Bruno, Jean-Paul Delahaye and Philippe Mathieu. "Complete Classes of Strategies for the Classical Iterated Prisoner's Dilemma" in *Evolutionary Programming, 1998*: 33 - 41.

[4] Binmore, Kenneth G. *Fun and Games: A Text on Game Theory*. Lexington, MA: D.C. Heath and Co. 1992.

[5] Brams, Steven J. *Theory of Moves*. Cambridge: Cambridge University Press. 1994.

[6] Cohen, Michael D., Rick L. Riolo and Robert Axelrod. *The Emergence of Social Organization in the Prisoners' Dilemma: How Context-Preservation and other Factors Promote Cooperation*. Santa Fe Institute Working Paper 99-01-002. 1998.

[7] Coxeter, Harold S.M and W.O.J. Moser. *Generators and Relations for Discrete Groups, 2e*. New York: Springer-Verlag. 1965.

[8] Eaton, B. Curtis. "The Elementary Economic of Social Dilemmas" in *Canadian Journal of Economics: November, 2004*.

[9] Flood, Merrill M. "Some Experimental Games" in *Research Memorandum RM-789*. Santa Monica, CA: RAND Corporation. 1952.

[10] Greenberg, Joseph. *The Theory of Social Situations: an Alternative Game Theoretic Approach*. Cambridge: Cambridge University Press. 1990.

[11] Hamburger, H. "Separable Games" in *Behavioural Science, 14*: 121 - 132. 1969.

[12] Hausdorff, Felix. *Grundzüge der Mengenlehre*. Leipzig: Verlag von Veit & Comp. 1914.

[13] Heylighen, F. "Game Theory" from *Principia Cybernetica Web*. (Eds) F. Heylighen, C. Joslyn, and V. Turchin. Brussells: Principia Cybernetica. http://pespmc1.vub.ac.be/ASC/GAME_THEOR.html (2000).

[14] Jones, Antonia J. *Game Theory: Mathematical Models of Conflict* (Horwood Series in Mathematics and Applications). Chichester: Horwood Publishing. 2000.

[15] King, Daniel. "Game Theory: The Study of Conflict and Strategy (Course Syllabus, Spring 2003)". (Sarah Lawrence College, Bronxville NY.) http://pages.slc.edu / dking/GameTheory_Syllabus.doc (2003).

[16] Licht, Amir N. "Games Commissions Play: $2 \times 2$ Games of International Securities Regulation" in *Yale Journal of International Law, Vol. 24 Number 1, Winter 1999*. New Haven, CT: Yale Journal of International Law, Inc.

[17] Luce, R. Duncan and Howard Raiffa. *Games and Decisions: Introduction and Critical Survey*. New York: John Wiley & Sons. 1967.

[18] Munkres, J. R. *Topology: A First Course*. Englewood Cliffs, NJ: Prentice-Hall. 1975.

[19] Myerson, Roger B. *Game Theory: Analysis of Conflict*. Cambridge, MA: Harvard University Press. 1997.

[20] Poundstone, William. *Prisoners's Dilemma: John von Neumann, Game Theory, and the Puzzle of the Bomb*. New York: Doubleday. 1992.

[21] Rapoport, Anatol. "Conflict Resolution in the Light of Game Theory and Beyond" in *The Structure of Conflict*. Ed. Paul Swingle. New York: Academic Press. 1970.

[22] Rapoport, Anatol. *Game Theory as a Theory of Conflict Resolution*. Kluwer Academic Publishers Group. 1974.

[23] Rapoport, Anatol and Melvyn J. Guyer. "A Taxonomy of $2 \times 2$ Games" in *General Systems, Vol. XXIII (1978)*: 125 - 136.

[24] Rapoport, Anatol, Melvyn J. Guyer and David D. Gordon. *The* $2 \times 2$ *Game*. Ann Arbour: University of Michigan Press. 1976.

[25] Robinson, D.R. and D.J. Goforth. "A Topologically-based Classification of the $2 \times 2$ Ordinal Games". Presented at the 37th Annual CEA Conference, Ottawa, May 30 to June 1, 2003. http://economics.ca/2003/papers/0439.pdf (2004).

[26] Robinson, D.R. and D.J. Goforth. "Graphs and Groups for the Ordinal $2 \times 2$ Games." Presented at the Canadian Economic Theory Conference, Montreal, May 7 - 9, 2004.

[27] Robinson, D.R. and D.J. Goforth. "Old Friends and Warped Relatives: The Search for the Periodic Table of $2 \times 2$ Games." Laurentian University, Economics Department Seminar, April, 2002.

[28] Robinson, D.R. and D.J. Goforth. "Can Game Theory Explain the Patriarchy?" Presented at the 36th Annual CEA Conference, Montreal, June 3 - 7, 2001.

[29] Robinson, D.R. and D.J. Goforth. "Reinterpreting the Repeated Prisoner's Dilemma: A Marginal Productivity Approach". Presented at the 35th Annual CEA Conference, Vancouver, June 1 - 4, 2000.

[30] Rosero, Alvaro Francisco Huertas. *Classification of Quantum Symmetric Nonzero-sum* $2 \times 2$ *Games in the Eisert Scheme*. Thesis presented at the Universidad de los Andes, Merida, Venezuela, February, 2004. http://arxiv.org/PS_cache/quant-ph/pdf/0402/0402117.pdf

[31] Sandholm, Tuomas and Robert Crites. "Multiagent Reinforcement Learning in the Iterated Prisoner's Dilemma" in *Biosystems, 37*: 147 - 166. 1995.

[32] Schelling, Thomas C. *Arms and Influence*. New Haven: Yale University Press. 1966.

[33] Schelling, Thomas C. *The Strategy of Conflict*. Oxford: Oxford University Press. 1963.

[34] Schelling, Thomas C. "Prospectus for a Re-orientation of Game Theory" in *Journal of Conflict Resolution, 2*: 203 - 264. 1958.

[35] Snyder, Glen H. and Paul Diesing. *Conflict Among Nations: Bargaining, Decision Making and System Structure in International Crises*. Princeton: Princeton University Press. 1971.

[36] Straffin, Philip D. Jr. "The Prisoner's Dilemma" in *UMAP Journal, 1*: 102 - 103. Lexington, MA. COMAP Inc. 1980.

[37] Swingle, Paul. "Dangerous Games" in *The Structure of Conflict*. Ed. Paul Swingle. New York: Academic Press. 1970.

[38] Ullmann-Margalit, Edna. *The Emergence of Norms*. Oxford. Oxford Univerity Press. 1977.

[39] von Neumann, John and Oskar Morgenstern. *Theory of Games and Economic Behavior*. Princeton: Princeton University Press. 1944.

[40] von Stengel, Bernhard. "Course Outline for MA301: Game Theory I, 2003 - 2004" (The London School of Economics and Political Science, London). http://www.maths.lse.ac.uk/Courses/ma301.html (2004).

[41] Weise, K.H. and H. Noack. "Aspects of Topology" in *Fundamentals of Mathematics, Volume II: Geometry* (Chapter 16). Eds. H. Behnke, F. Bachmann and K. Fladt. Cambridge, MA: MIT Press. 1986. English translation of *Grundzüge der Mathematik* (second German edition, 1967).

[42] Weisstein, Eric W. "Hausdorff Axioms." From *MathWorld: A Wolfram Web Resource*. http://mathworld.wolfram.com/HausdorffAxioms.html (2004).

[43] Weisstein, Eric W. "Villarceau Circles." From *MathWorld: A Wolfram Web Resource*. http://mathworld.wolfram.com/VillcarceauCircles.html (2004).

# Index